Identifying and Prioritizing Emerging Safety Issues in the Water Industry

The mission of the Awwa Research Foundation (AwwaRF) is to advance the science of water to improve the quality of life. Funded primarily through annual subscription payments from over 1,000 utilities, consulting firms, and manufacturers in North America and abroad, AwwaRF sponsors research on all aspects of drinking water, including supply and resources, treatment, monitoring and analysis, distribution, management, and health effects.

From its headquarters in Denver, Colorado, the AwwaRF staff directs and supports the efforts of over 500 volunteers, who are the heart of the research program. These volunteers, serving on various boards and committees, use their expertise to select and monitor research studies to benefit the entire drinking water community.

Research findings are disseminated through a number of technology transfer activities, including research reports, conferences, videotape summaries, and periodicals.

Identifying and Prioritizing Emerging Safety Issues in the Water Industry

Prepared by:
Peter S. Puglionesi, **David Baird**, and **Shafat Shaon**
Duke Engineering and Services
1800 Chapel Avenue West, Suite 300
Cherry Hill, NJ 08002

Sponsored by:
Awwa Research Foundation
6666 West Quincy Avenue
Denver, CO 80235-3098

Published by the

Awwa Research Foundation
Advancing the Science of Water®

IWA Publishing

DISCLAIMER

This study was funded by the Awwa Research Foundation (AwwaRF). AwwaRF assumes no responsibility for the content of the research study reported in this publication or for the opinions or statements of fact expressed in the report. The mention of trade names for commercial products does not represent or imply the approval or endorsement of AwwaRF. This report is presented solely for informational purposes.

Published by IWA Publishing, Alliance House, 12 Caxton Street, London SW1H 0QS, UK

Telephone: +44 (0) 20 7654 5500; Fax: +44 (0) 20 7654 5555; Email: publications@iwap.co.uk;
Web: **www.iwapublishing.com**

AwwaRF report number 90935F. Originally published by AwwaRF for its subscribers in 2003.
IWA Publishing version published 2004.

ISBN 1 84339 844 3

CONTENTS

LIST OF TABLES

LIST OF FIGURES

FOREWORD

The Awwa Research Foundation is a nonprofit corporation that is dedicated to the implementation of a research effort to help utilities respond to regulatory requirements and traditional high-priority concerns of the industry. The research agenda is developed through a process of consultation with subscribers and drinking water professionals. Under the umbrella of a Strategic Research Plan, the Research Advisory Council prioritizes the suggested projects based upon current and future needs, applicability, and past work; the recommendations are forwarded to the Board of Trustees for final selection. The foundation also sponsors research projects through the unsolicited proposal process; the Collaborative Research, Research Applications, and Tailored Collaboration programs; and various joint research efforts with organizations such as the U.S. Environmental Protection Agency, the U.S. Bureau of Reclamation, and the Association of California Water Agencies.

This publication is a result of one of these sponsored studies, and it is hoped that its findings will be applied in communities throughout the world. The following report serves not only as a means of communicating the results of the water industry's centralized research program but also as a tool to enlist the further support of the nonmember utilities and individuals.

Projects are managed closely from their inception to the final report by the foundation's staff and large cadre of volunteers who willingly contribute their time and expertise. The foundation serves a planning and management function and awards contracts to other institutions such as water utilities, universities, and engineering firms. The funding for this research effort comes primarily from the Subscription Program, through which water utilities subscribe to the research program and make an annual payment proportionate to the volume of water they deliver and consultants and manufacturers subscribe based on their annual billings. The program offers a cost-effective and fair method for funding research in the public interest.

A broad spectrum of water supply issues is addressed by the foundation's research agenda: resources, treatment and operations, distribution and storage, water quality and analysis, toxicology, economics, and management. The ultimate purpose of the coordinated effort is to assist water suppliers to provide the highest possible quality of water economically and reliably. The true benefits are realized when the results are implemented at the utility level. The foundation's trustees are pleased to offer this publication as a contribution toward that end.

This report seeks to enhance understanding of the frequency and the different types of injuries that occur in the water industry as a basis to identify emerging health and safety issues as well as future health and safety research needs for the water industry.

Edmund G. Archuleta, P.E.
Chair, Board of Trustees
Awwa Research Foundation

James F. Manwaring, P.E.
Executive Director
Awwa Research Foundation

ACKNOWLEDGMENTS

The authors of this report are indebted to a broad range of individuals, utilities, and other stakeholders for their cooperation and participation in this project. In this regard, we wish to acknowledge the contributions of the AwwaRF Project Manager, Robert Allen.

Volunteer Utilities

We wish to acknowledge the following utilities that committed significant resources and effort to respond to the survey, review the workshop materials, attend the workshop and review the draft project scopes and report.

- American Water Works Co., Inc., Chula Vista, California, Dennis Thorig
- Beaufort Jasper Water & Sewage Authority, Beaufort, South Carolina, Ken Jordan
- Borough of Carlisle, Carlisle, Pennsylvania, Mike McFadden
- Chester Water Authority, Chester, Pennsylvania, Thoedore J. Pawlik
- City of Denton, Denton, Texas, Byron Piechowski
- City of Fort Worth Water Department, Fort Worth, Texas, Gerald Keeling
- City of Phoenix Water Services Dept., Phoenix, Arizona, Mel Meevis
- City of Portland, Portland, Oregon, Eric J. Fullan and Clint Van Arsdall
- City of Raleigh Public Utilities Dept., Raleigh, North Carolina, David W. Massenburg
- City of Talahassee, Tallahassee, Florida, Joe Murphy
- Columbus Public Utilities Division of Water, Columbus, Ohio, Thomas G. Finnegan
- Durham Water Resources, Durham, North Carolina, Renee Lawrence
- Elizabeth Town Water Co., Westfield, New Jersey, Janis Morelli
- Fayetteville Public Works Commission, Fayetteville, North Carolina, Dave Holloway
- Los Angeles Dept. of Water and Power, Los Angeles, California, Leland Gong
- Louisville Water Company, Louisville, Kentucky, Glenn Mudd
- Memphis Light, Gas & Water, Memphis, Tennessee, Carolyn Turman
- Missouri-American Water Co., St. Louis, Missouri, Tony Dicandia
- Philadelphia Water Dept., Philadelphia, Pennsylvania, James D'Agostino, Jr.
- San Antonio Water System, San Antonio, Texas, Ben Skidmore
- United Water, Indianapolis, Indiana, Johanna Krachinski

Other Participaitng Stakeholders

We wish to acknowledge the following organizations for their participation as in-kind contributors of both time and reference information in providing data and presentations and their extensive participation the workshop and document review process:

- Bureau of Labor Statistics, Katharine Newman
- Duke Energy Corporation, Charlotte, North Carolina, Jeffrey Almond
- Florida Workers Compensation Division, Bobbie Markiewicz.
- National Institute of Occupational Safety and Health, Max Lum
- National Safety Council, Chicago, Illinois, Alan F. Hoskin.
- NSC Utility Division Chair/Consultant, Jefferson City, Missouri, Randy Brown.
- NY State Dept. of Public Employee Safety and Health, Jonathan Haney

AwwaRF Project Advisory Committee Members

We also thank the following water utilities and individuals who served as members of the Project Advisory Committee and very active participants in the project and workshop:

- Cincinnati Water Works, Cincinnati, Ohio, William Phelps
- North Marin Water District, Novato, California, Gayle Smalley

EXECUTIVE SUMMARY

RESEARCH OBJECTIVES

Water supply systems around the world are subject to the effects of work-related accidents and environmental health-related illnesses. Some organizations have been deeply affected by fatal incidents. Even "routine" injuries and illnesses affecting all organizations can result in lost productivity, increased worker's compensation cost, and poor employee morale.

The primary goals of this project are to identify emerging health and safety issues and identify future health and safety research needs for the water industry. It's other objectives include identifying potential research partners and enhancing understanding the type and frequency of injuries that occur in the water industry as a basis for utilities to develop and implement more effective safety programs.

This project can yield the following benefits to water utilities:

- Promoting more uniform injury/illness tracking can result in an improved ability to benchmark across the industry.
- Identifying divisions and job classifications/activities that have higher work-related injury and illness rates can result in better targeting training programs and hazard mitigation.
- Identifying conventional and state-of-the-art safety management approaches that significantly enhance performance and reduce cost of injuries can assist utilities in applying the most effective methods.
- Prioritizing injury/illness related safety issues that warrant greater research attention can help identify potential research projects that address the most critical safety issues.
- Prioritizing research projects based upon the criticality of the needs, the likelihood of success and the likely impact on performance improvement across the industry can result in more effective use of limited research dollars.

APPROACH

The project approach was based on bringing together the accumulated experience of Framatome ANP DE&S (F-DE&S) professional staff, those in the water industry, and other critical stakeholders who's mission is consistent with promoting reductions in workplace injuries and illness. A wide range of stakeholders were represented to ensure that research agenda documented in the final publication is broadly meaningful, useful and will have positive impacts for the industry as a whole, including utilities that did participate. The steps taken to meet the objectives of this project were as follows:

- Analyze information on the incidence of injuries and illnesses in the water industry collected by the AWWA Health and Safety Committee and other sources (e.g., the Bureau of Labor Statistics and National Safety Council).

- Develop and distribute a health and safety survey to collect further information on injury/illness data collection practices and successful approaches to safety performance improvement.
- Evaluate health and safety survey responses.
- Hold a workshop for water utilities and other stakeholders to review the results of the survey and the analysis of current injury/illness information, identify critical issues, and develop the research agenda.

CONCLUSIONS

Bureau of Labor Statistics data indicate that the water industry as a whole has higher recordable incidence rates, lost workday case rates, and severity rates[1] when compared to other utilities (i.e., electric utilities). In addition, government operated water utilities operate in an environment without health and safety regulations comparable to U.S. OSHA rules in almost half of the states of the U.S. It may be that these utilities are more susceptible to complacency and poor performance because regulatory enforcement is often the most potent factor driving health and safety programs.

Current performance suggests that water utility executives and industry associations such as AWWA and AwwaRF should make concerted efforts to improve health and safety performance. An R&D agenda is essential to direct available resources to the most beneficial projects and to provide a basis for receiving co-funding from other entities interested in the same goals. Because R&D funding is limited, this agenda (based upon the most critical needs in the industry and the areas that can most benefit from research) will yield the most benefits for the industry.

While AwwaRF can assist in furthering the goal of reducing injuries and illnesses in the industry by conducting research and development to provide long-term support of performance improvements, water utilities need not wait for AwwaRF research. We urge water utilities to make a commitment to compliance and safety excellence and to make aggressive use of existing safety management tools and techniques. This report presents a brief summary of lessons learned from the workshop and the contributions of participating health and safety specialists that can be useful to water utilities in improving their safety programs and performance.

The project workshop participants identified the most critical health and safety needs in the water industry. The top three health and safety needs identified are:

- Training
- Management commitment
- Formal written health and safety programs

If resources are to be allocated based upon the incidence and severity rates, water utilities should prioritize their efforts and resources as follows:

[1] This is the incidence rate of lost workdays as defined by BLS and OSHA (also called Lost Workday Incidence Rate).

Highest Priority:
- Distribution Maintenance
- Surface Water Treatment
- Meter

Second Priority:
- Source of Supply
- Distribution Construction
- Maintenance

Third Priority:
- Ground Water Treatment
- Administration
- Laboratory

RECOMMENDATIONS FOR FUTURE RESEARCH

Eight critical research projects were identified by the participants in this workshop and prioritized based upon both need and likelihood of success in achieving safety performance improvement, as described in Section 4. These projects, in order of priority, are:

1. Effectiveness of Techniques for Prevention of Lower Back and Shoulder Injuries and Other Musculoskeletal Disorders
2. Best Practices for Integrating Health and Safety into the Design and Operation of Water Systems
3. Establishing the Full Costs of Injuries and Illnesses and Evaluating the Cost/Benefit of Prevention Programs
4. National Water Occupational Health & Safety Database: Feasibility and Pilot of Benchmarking Data Consortium
5. Evaluation of the Effectiveness of Alternate Enforcement Methods and Approaches
6. Effective Infrastructure Protection Techniques for Protection of Water System Employees and the Public
7. The Role of Human Factors in Injuries and Improving Safety Performance in the Water Industry
8. The Effects of Chronic Low Level Exposures to Airborne Chemicals at Water Utilities

AwwaRF research and development is devoted to achieving a wide variety of objectives and funding for any one of these objectives is limited. The workshop participants recommend that the goal of reducing injuries/illnesses be among the top AwwaRF priorities and that a minimum of 1-2 health and safety projects be funded in each year. With that level of AwwaRF commitment, the above list represents an R&D agenda for the 5-10 year horizon.

CHAPTER 1
INTRODUCTION

Water supply systems around the world are subject to the effects of work-related accidents and environmental health-related illnesses. Some organizations have been deeply affected by fatal incidents. Even “routine” injuries and illnesses affecting all organizations can result in lost productivity, increased worker's compensation cost, and poor employee morale.

This project report documents efforts to identify emerging health and safety issues and future health and safety research needs for the water industry. This report also serves to inform water industry executives and Health and Safety staff regarding the type and frequency of injuries that occur in the water industry. This can serve as a basis for utilities to develop and implement more effective and well directed safety programs.

Those who are mainly interested in using the results of this work for improving their Health and Safety programs are encouraged to skip ahead to Section 3.

CURRENT STATE OF KNOWLEDGE

Numerous organizations are interested in preventing on-the-job injuries and illnesses (e.g., insurance companies, research institutions, government agencies and non-profit professional and industry associations), but the focus of each of these organizations varies.

AWWA, for instance has conducted an annual safety survey on water industry accident performance as part of a safety award program. The insurance industry collects data only for organizations they provide with insurance coverage or providc self-insurance administration services. The U.S. Department of Labor Bureau of Labor Statistics (BLS) collects extensive injury/illness data, but only for privately operated water utilities. For publicly owned water utilities, data collection is highly fractured and based upon widely varying State government agency rules.

Because publicly operated water systems are not uniformly reporting data, the current accident/injury/illness data collection and reporting practices in the water industry tends to vary widely. For example:

- Privately owned and/or operated systems at minimum report accidents and injuries in conformance with Occupational Safety and Health Administration (OSHA) requirements.
- Publicly owned and operated systems are exempt from Federal OSHA regulations based on the original enabling legislation; however, in OSHA Plan States or States that have special requirements for government employers, these systems report to State labor safety agencies in conformance with Federal OSHA requirements.
- Systems that carry workman's compensation insurance or are self-insured and use insurance company administrative support may use insurance company reporting practices and databases.
- Systems that are self-insured and self-administered may have to meet the data requirements of the particular State's workman's compensation laws.

- Some organizations with a pro-active safety management approach may have added significantly to the injury/illness reporting system to include recording and evaluating near misses, including extensive reporting on causal factors, and/or including root cause analysis as part of the investigation protocols.

In addition to the factors cited above, the injury/illness experiences in certain organizations may also be affected by traditional labor agreements and labor/management relationships. Inherent differences in business and work activities from organization to organization adds a separate layer of complexity to the already non-uniform reporting practices making it difficult for organizations to benchmark their loss experiences. For example:

- Systems that are extensively involved in collection, treatment and distribution of water for wholesaling will have a reduced portion of worker hours spent on water distribution system maintenance and meter reading.
- Systems that buy treated water may have a disproportionately high distribution system maintenance and meter reading component.
- Very new systems will have much lower distribution system maintenance labor as compared with very old systems.
- Meter reading practices vary widely from monthly to quarterly to using automatic telemetry to contracting meter reading. Meter locations may vary from in-home in the north to outside of home in the south.

OBJECTIVES

The primary goals of this project are to identify emerging health and safety issues and identify future health and safety research needs for the water industry. Objectives include identifying potential research partners, and enhancing the understanding the type and frequency of injuries that occur in the water industry as a basis for utilities to develop and implement more effective safety programs.

This project can yield the following benefits to water utilities:

- Promoting more uniform injury/illness tracking can result in an improved ability to benchmark across the industry.
- Identifying divisions and job classifications/activities that have higher work-related injury and illness rates can result in better targeting training programs and hazard mitigation.
- Identifying conventional and state-of-the-art safety management approaches that significantly enhance performance and reduce cost of injuries can assist utilities in applying the most effective methods.
- Prioritizing injury/illness related safety issues that warrant greater research attention can help identify potential research projects that address the most critical safety issues.
- Prioritizing research projects based upon the criticality of the needs, the likelihood of success and the likely impact on performance improvement across the industry can result in more effective use of limited research dollars.

CHAPTER 2
TECHNICAL APPROACH

The project approach was based on bringing together the accumulated experience of Framatome ANP DE&S (F-DE&S) professional staff, those in the water industry, and other critical stakeholders who's mission is consistent with promoting reductions in workplace injuries and illness. A wide range of stakeholders were represented to ensure that research agenda documented in the final publication is broadly meaningful, useful and will have positive impacts for the industry as a whole, including other utilities not participating..

The steps taken to meet the objectives of this project were as follows:

- Analyze information on the incidence of injuries and illnesses in the water industry collected by the AWWA Health and Safety Committee and other sources (e.g., the Bureau of Labor Statistics and National Safety Council).
- Develop and distribute a health and safety survey to collect further information on injury/illness data collection practices and successful approaches to safety performance improvement.
- Evaluate health and safety survey responses.
- Hold a workshop for water utilities and other stakeholders to review the results of the survey and the analysis of current injury/illness information, identify critical issues, and develop the research agenda.

The following sections describe these steps in more detail.

DATA COLLECTION AND ANALYSIS

F-DE&S initiated this project by reviewing the accident/injury data already accumulated by the AWWA Health & Safety Committee for a variety of water utilities. Preliminary information on accident/injury data management systems, data structures, health and safety management practices and safety concerns/issues were collected from participating water utilities and insurance companies other potential industry participants that represent a wide range of geographic locations and type of operations. A questionnaire was developed and included the following considerations:

- Organization characteristics (capacity, number of employees in treatment/distribution/ customer service/etc., number of customers, type of facilities, public/private ownership, self/other insured, self-administered/contract-administered, etc.).
- Reporting and classification systems for injury, occupational illness and personnel data (including whether there is cost tracking, breakdown by operating division/job classification/ etc., identification of causal factors/root causes, reporting of first aid cases/near misses, etc.)
- Availability, form and accessibility of current and historical data on injury/illness.
- Availability of records for total work hours by division and job classification and methods of compiling work hours.

- Current health and safety programs and practices for compliance and safety performance improvement.
- Management and employee perceptions about current health and safety programs, their effectiveness, their importance, the key areas that require improvement, and their opinions about how research could improve safety performance in the industry.
- Other health and safety data (e.g., workplace or medical monitoring data).

The responses from the questionnaire were reviewed with respect to the potential data usability for more in-depth evaluation to help identify safety performance problem areas and issues. This evaluation included a comparison of performance between the different operating divisions and job function/classifications, identification of more prevalent injuries and identification of work activities/job factors that may result in higher incidence rates.

HEALTH AND SAFETY WORKSHOP

After evaluating the survey responses, a 2-day workshop was conducted to review the injury/illness data evaluation and results of the survey, identify critical safety issues and develop an agenda for future research. The workshop discussions were prompted and initiated by various presentations followed by facilitator led discussions and work product development.

Workshop Presentations

The Workshop included presentations that cover both the industry survey results and detailed injury/illness data evaluation and key issues pre-identified by the survey respondents and workshop participants as critical or emerging issues.

Facilitated Brainstorming Process

A structured brainstorming process was held for identifying issues and potential solutions, prioritizing the issues/solutions and developing work scope descriptions for the highest ranking potential research projects.

Developing Research Project Descriptions

After the step-wise identification of needs, potential solutions, necessary research projects and priority from Workshop participants, the workshop developed brief research project summaries for the top research priority areas. Breakout groups developed "bulleted" project scope descriptions addressing the key research project scope categories of Objective, Background, Research Approach and Estimated Budget. The breakout groups also identified potential research partners who could potentially share in the funding and/or execution of the research.

CHAPTER 3
WATER INDUSTRY SAFETY PERFORMANCE

EXISITING SOURCES

1999 AWWA Safety Survey

The AWWA Health and Safety Committee provided responses to the 1999 safety survey to serve as the basis for this project. The survey was designed by the Health and Safety Committee and administered by AWWA staff. AWWA staff cautioned against considering the data statistically valid because of the following factors:

- The intent of the survey was to make AWWA safety awards, not to collect fully verified data representative of the industry.
- AWWA identified clear inconsistencies in numerous responses and tried to resolve them.
- Many respondents were from very small water systems that did not have safety staff and often did not have a clear understanding of the common definitions used (e.g., recordable incidences, lost workday, etc.).
- Due to staff and time limitations, not all apparent problems could be resolved.
- Approximately 6000 surveys are sent and over 1350 responses were received. Since the responses determine AWWA safety awards, responses could potentially be skewed toward those who believe they have better safety records (although performance does not appear better than other, more qualified performance data).

Due to these data limitations, AWWA has requested that the safety statistics derived from the survey not be distributed or cited beyond the AwwaRF Project Participants.

Because of these limitations, absolute safety statistics from this survey are not presented. This data was only used to compare the relative importance of potentially critical factors affecting safety performance to support identifying the greatest needs for Health and Safety research in the industry.

National Safety Council Data

Each year, the National Safety Council (NSC) reports safety data for each industry group in an annual report. NSC data are compiled only from Council members that chose to participate in the NSC's *Occupational Safety/Health Award Program.* As a result, NSC cautions that "because participation is voluntary, the resulting incidence rates are not representative of the industries or of Council members." It is also likely that NSC members from the water industry are more likely to be larger organizations with greater resources to participate in safety organization activities than the smaller organizations that comprise a large number of AWWA members and survey responders.

This data was used for comparison with other nationally compiled data sources. It was also used to compare performance between utility industry groups (i.e., water, wastewater, power and combined utilities).

Bureau of Labor Statistics Data

The Bureau of Labor Statistics reports annual safety incidence data compiled through a nationwide random sample of establishments for each industry group. The sample is selected to be representative of all private employers in the United States.

The BLS data is collected in a rigorous manner and is considered to be representative of each industry; however, for the water industry the privately owned and operated utilities are just one segment of the industry. The BLS data is not representative of the publicly owned and operated segment. Data including public systems is only available on a state-by-state basis for those states that have safety regulations applying to municipalities. This may be a significant distinguishing factor, in that BLS data is collected from organizations that are subject to OSHA requirements. These organizations may perform better due to the greater regulatory requirements, enforcement and the potential for a resulting higher average safety programs effort.

This data was also used for comparison with other nationally compiled data sources. The BLS data, unlike AWWA and NSC data does not include a Severity Rate.

EVALUATION OF EXISTING DATA SOURCES

The available information on the incidence of injuries and illnesses in the water industry collected by the AWWA Health and Safety Committee, NSC and by BLS allowed us to evaluate the data in several ways, as follows:

- Recordable Rate, Lost Workday Case Rate and Severity Rate (see Appendix A for definitions) was compared between the different types of utilities (i.e., water only, water and wastewater and combined with power and/or natural gas) using the NSC and BLS data sources.
- The AWWA data was sorted into facilities in OSHA Plan states and those in other states to determine if there are apparent performance differences between OSHA-regulated and non-OSHA regulated utilities.

The reviewer should be aware that the BLS data does not represent publicly owned and operated systems. An organization in both the water and wastewater business would fall into the NSC and BLS "Combined Utilities" category, as would an organization that combined electric power service with water and/or wastewater.

The attributes of these data sources can be summarized as follows:

Table 3.1
Advantages and Disadvantages of Existing Data Sources

Data Source	Advantages	Disadvantages
AWWA Safety Survey	▪ Largest number of water industry respondents ▪ Excellent representation of public utilities	▪ Limited data quality verification ▪ No statistical analysis ▪ Voluntary participation may bias response toward those who have good performance in a given year
National Safety Council	▪ Highly knowledgeable members suggest good data quality	▪ Limited number of water industry respondents ▪ Respondents are members of NSC, a self-selected higher performing group
Bureau of Labor Statistics	▪ Random selection, highly representative of private water industry ▪ Excellent data quality verification and statistical analysis	▪ No public water systems included ▪ No severity rate (lost workday incidence rate)

Comparison of Performance Between Industries and From Different Data Sources

Figure 3.1 through Figure 3.3 present the Recordable Incidence Rate, Lost Workday Case Rate and Severity Rate for the various utility industry groups and combinations of utility groups discussed above for the BLS and NSC data sources. The following comparisons were made:

Comparison of Water Industry Rates to Wastewater Utilities

Comparisons of "Water Only" utility data to "Wastewater" utilities in Figure 3.1 through Figure 3.3 yields mixed results, depending upon which rate is compared. "Water" Recordable Incidence, Lost Workday Case and Severity rates are similar to the rates for "Sanitary Services[2]" in the NSC data. Recordable Incidence and Lost Workday Case rates for "Sanitary Services" are significantly higher in the BLS data. Although the BLS data would normally be considered more representative, it would seem that the exclusion of municipal systems from the data would leave few organizations representative of the wastewater industry. Accordingly, it is difficult to make any hypotheses based on this comparison.

[2] Note: Sanitary Services (SIC Code 495) includes sewerage systems and solid waste or refuse systems.

Comparison of Water Industry Rates to Electric Utilities

Comparisons of BLS and NSC "Water Only" utility data to "Electric" utilities in Figure 3.1 through Figure 3.3 indicate that there are consistently lower rates of recordable incidence and lost time in Electric utilities. Comparisons of "Water Only" utility data to "Combined" utilities in Figure 3.1 through Figure 3.3 indicate that there are consistently lower rates of incidence and lost time in "Combined" utilities in all data sets. The BLS statistics indicate that "Combined" utility performance falls in between "Electric" and "Water" utility performance while NSC statistics indicate that performance is similar between "Combined" and "Electric" these categories. (Note: some NSC and BLS combined utilities may include electric/gas combinations, but a significant number are likely to also include water and wastewater operations.) These data lead to the following observations and potential hypotheses:

- Electric utilities have significantly lower incidence and severity rates than water utilities.
- "Combined" utilities incidence and severity rates are also significantly lower than for water utilities, but not necessarily as low as Electric utilities.
- The lower incidence and severity rates for the Electric and Combined utilities leads to the following potential hypotheses for explaining the differences in rates:
  - Many work hazards at water utilities are inherently different from Electric utilities, although there does not appear to be a greater degree of inherent hazard associated with Water utility work.
  - Safety culture may be stronger and resource levels may be higher in the Electric Utilities, potentially due to the higher inherent hazard of high voltage and/or the average larger size of the organizations.
  - Rates may be higher for public water systems in over 20 states that are not regulated under federal OSHA regulations (see analysis in later section of OSHA Plan vs. non-Plan States).

The differences in types of work and associated hazards is not believed to solely account for higher incidence rates, but this is a complex factor. Electric utilities do more construction and repair of overhead lines involving higher hazard activities (e.g., work at heights/falls, falling objects, cranes, high voltage, etc.). Electricity is more inherently hazardous than water, but water utilities do more construction and repair of underground lines, involving activities (i.e., hand digging, tight spaces, manual lifting, potential for cave in, etc.) that may have higher injury frequencies.

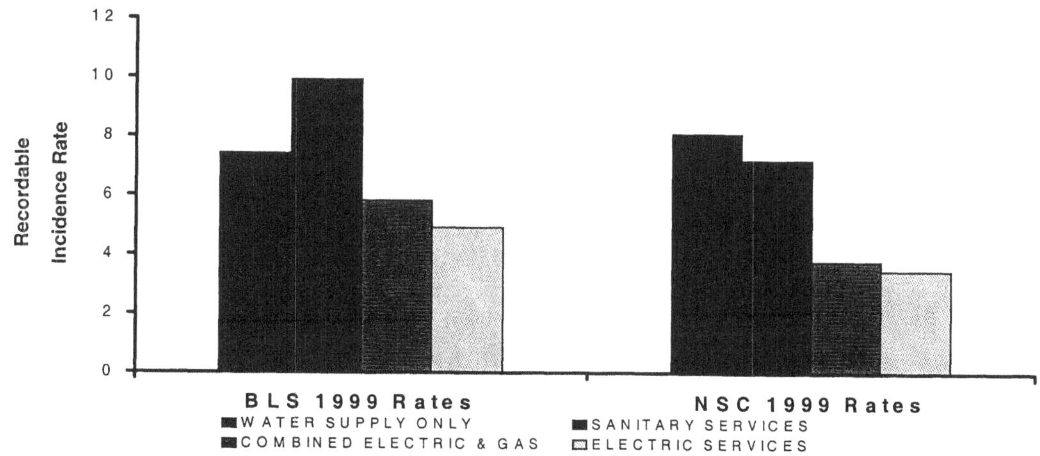

Figure 3.1 Recordable Incidence Rates (*Source*: BLS, 2000; NSC, 2000)

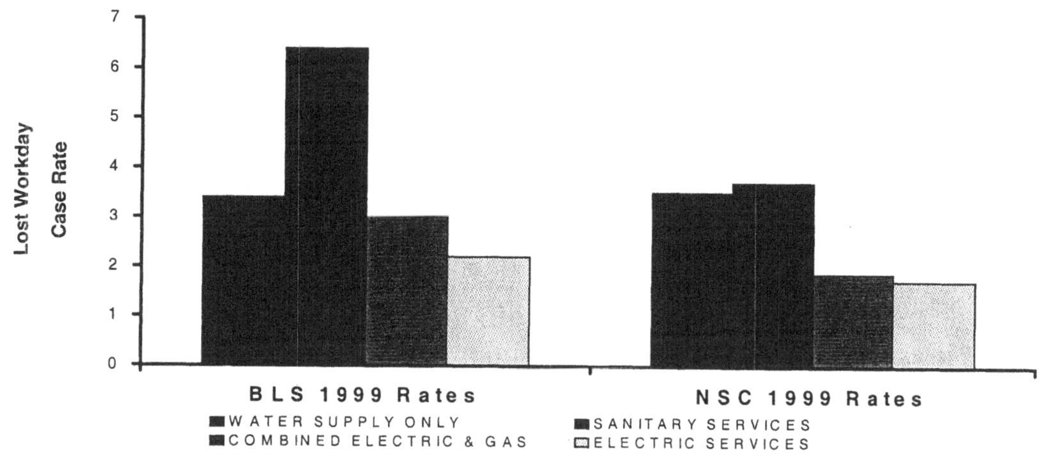

Figure 3.2 Lost Workday Case Rates vs. type of utility (*Source*: BLS, 2000; NSC, 2000)

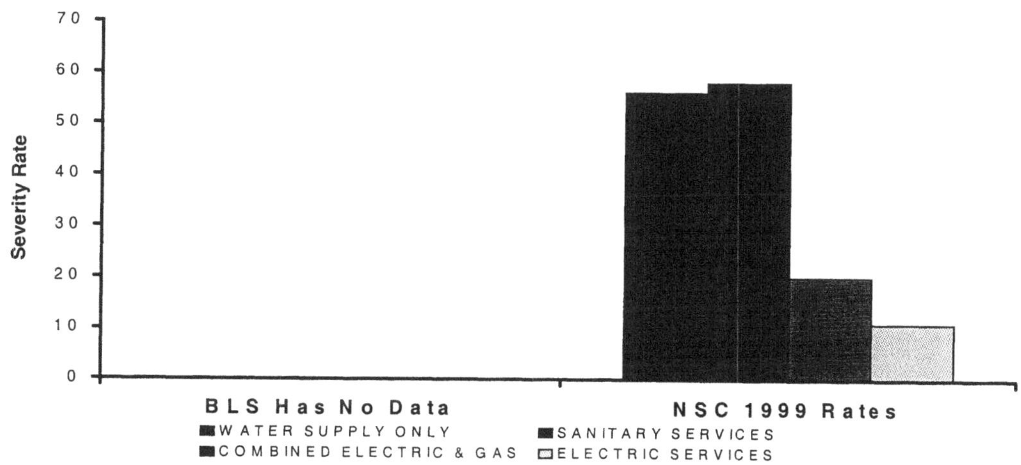

Figure 3.3 Severity Rates vs. type of utility (*Source*: BLS, 2000; NSC, 2000)

Comparison of Performance for OSHA Plan and Non-Plan States

The AWWA Safety Survey data was sorted into facilities located in OSHA Plan States (see Figure 3.4) and those located in other states to help evaluate whether the above hypothesis that non-OSHA regulated utilities' performance may be significantly different from OSHA regulated utilities. The data did not allow us to further separate out the OSHA regulated privately operated systems in OSHA Plan states.

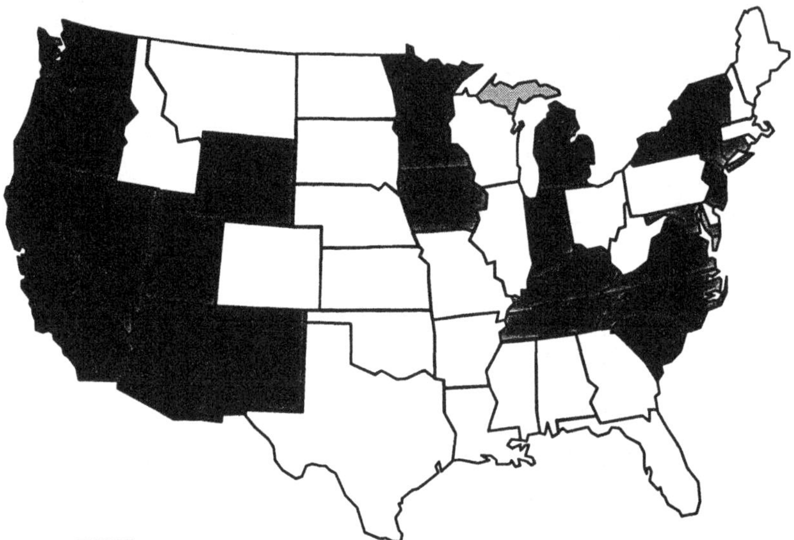

■ - OSHA Plan State (States Authorized to Enforce U.S. OSHA Rules and Apply Them to Government Employers)

Also: Alaska, Hawaii, Puerto Rico, Virgin Islands

Figure 3.4 OSHA Plan states

Figure 3.5 shows the percentage of survey respondents located in OSHA Plan and non-Plan states. Figure 3.6 and Figure 3.7 indicate that Recordable Incidence Rates and Lost Workday Case Rates are more than 20% higher for non-OSHA Plan states. Figure 3.8 indicates that there is also a difference in Severity Rates, but this difference is smaller than the difference in Incidence Rates.

These results appear to confirm the hypothesis proposed in Section 0 that not being regulated under OSHA or OSHA equivalent regulations has a negative effect on safety performance. The most likely root cause is the absence of regulatory requirements and enforcement and the potential effect on safety program effort at public utilities.

Comparison of Number of Incidences in Different Departments

Conventional wisdom in the water industry is that the distribution system/construction function involves the most hazardous work and has the highest incidence rates. The AWWA Safety Survey obtained incidence data for each typical type of water industry department/operation.

Figure 3.9 and Figure 3.10 present the percent of total Recordable Cases with NO Lost Workdays and WITH Lost Workdays, respectively, experienced by each department. Figure 3.11 presents the percent of Total Lost Workdays experienced by each department. This data indicates that, as was expected, the distribution/construction organizations are experiencing a higher number of incidents and more severe incidents. Several other departments appear to be in a second tier with the priority order changing depending upon which parameter is used. This would appear to justify focusing effort and resources in the following order:

Highest Priority:

- Distribution/Construction

Second Priority:

- Meter or Customer Service
- Water Treatment
- Maintenance
- Source of Supply

Third Priority:

- Administration
- Laboratory

This is not to suggest that safety programs are unimportant in the lower priority areas. The reader should also be aware that the percentage distributions shown are based upon the total number of incidences and lost time days for each department, and not the incidence rates. The absolute number of cases and lost workdays are higher for inherently larger departments while true "rates" are normalized for the number of employee hours worked. This limitation was addressed in the survey conducted by F-DE&S, as discussed in the next section.

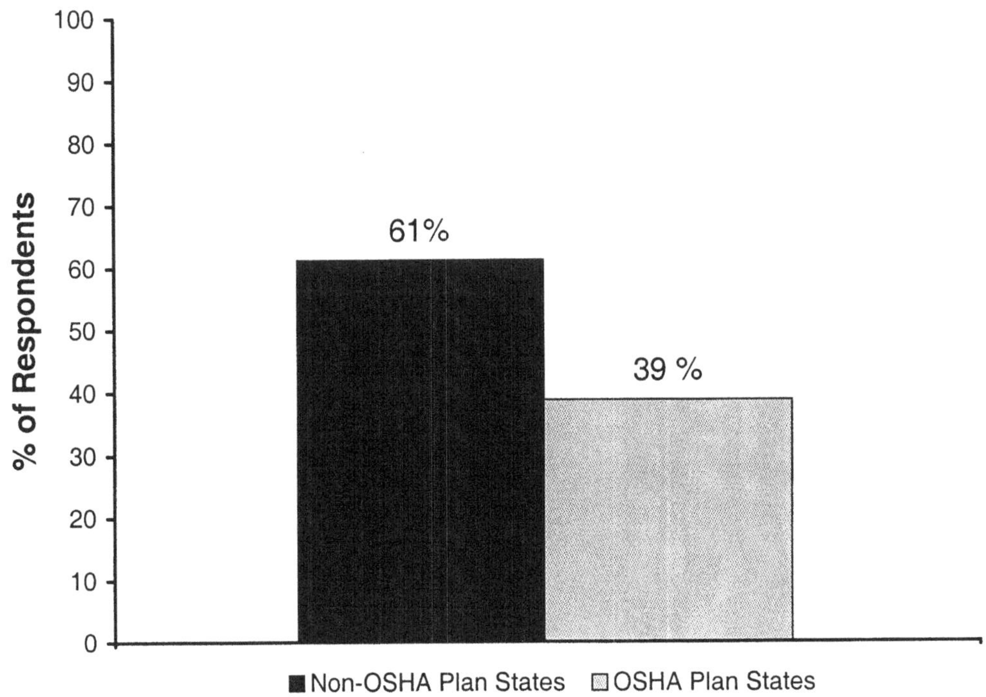

Figure 3.5 Percentage of 1999 AWWA survey respondents in OSHA-Plan and "Non OSHA-Plan" states (*Source*: Data from AWWA, 1999)

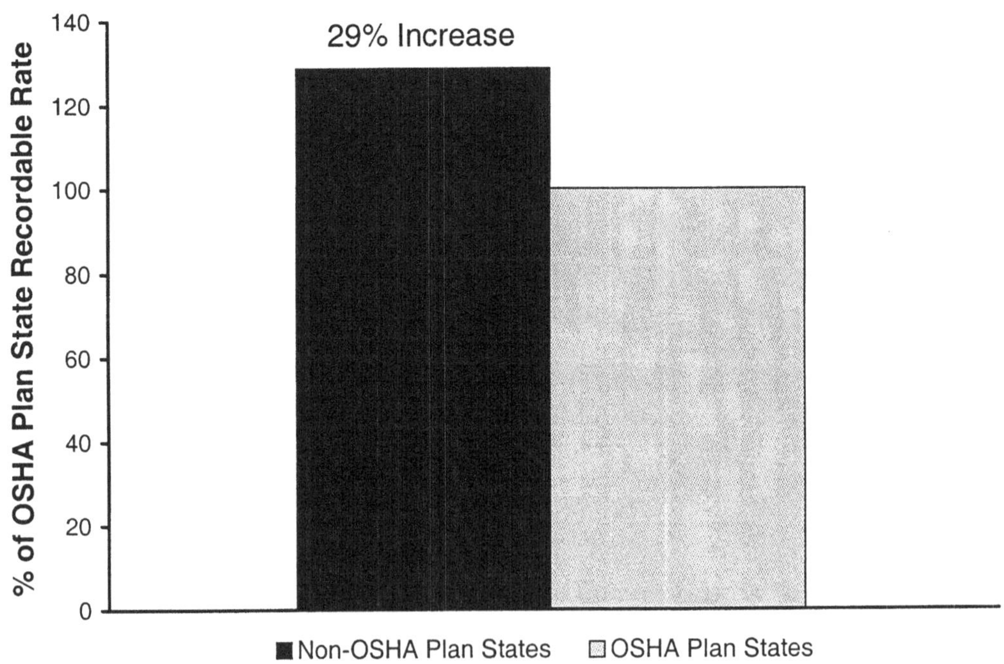

Figure 3.6 Comparison of 1999 water industry Recordable Incidence Rate between OSHA/Non-OSHA states (*Source*: Data from AWWA, 1999)

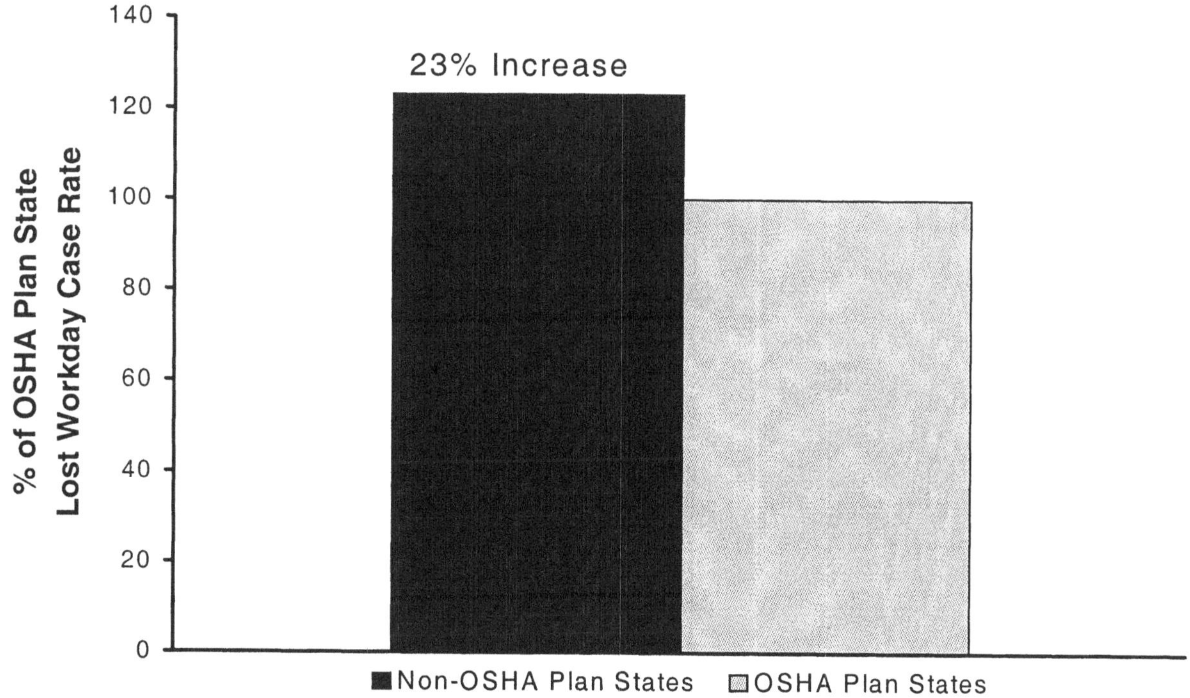

Figure 3.7 Comparison of 1999 water industry Lost Workday Case Rate between OSHA/Non-OSHA states (*Source*: Data from AWWA, 1999)

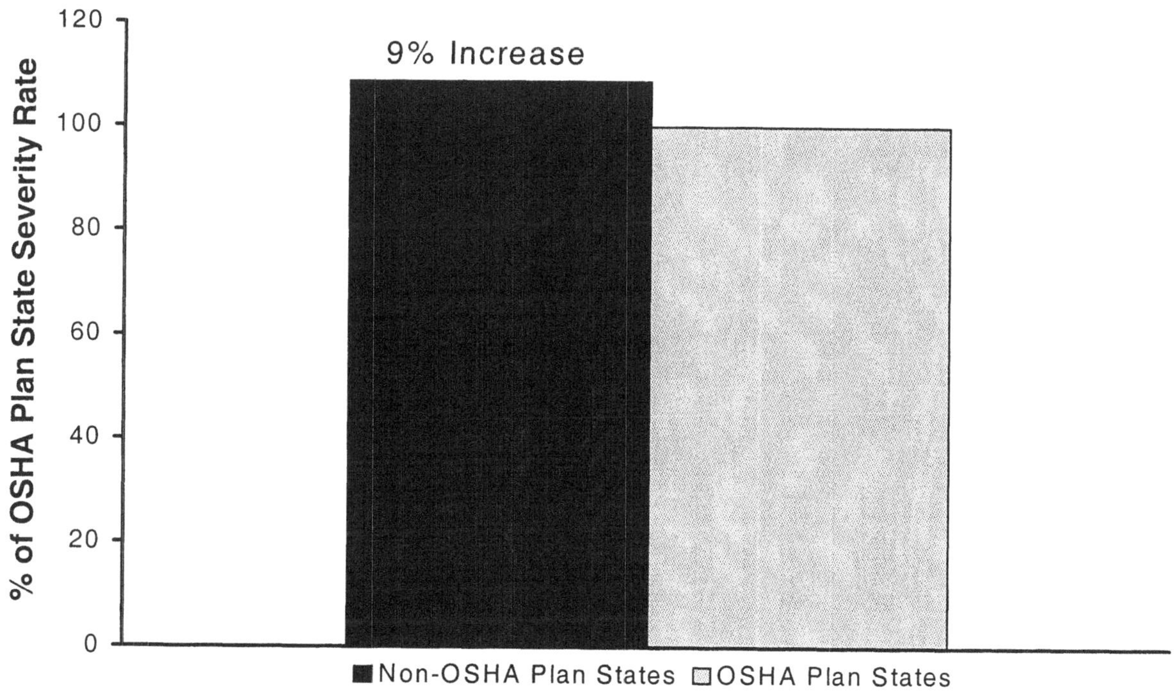

Figure 3.8 Comparison of 1999 water industry Severity Rate between OSHA/Non-OSHA States (*Source*: Data from AWWA, 1999)

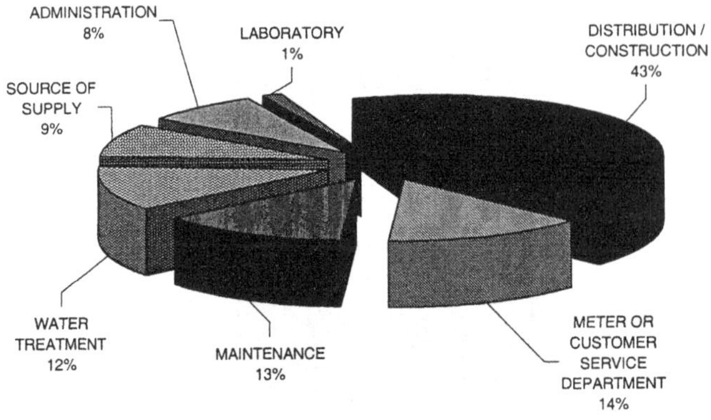

Figure 3.9 Percent Recordable Cases with no Lost Workdays for each type of operation (*Source*: Data from AWWA, 1999)

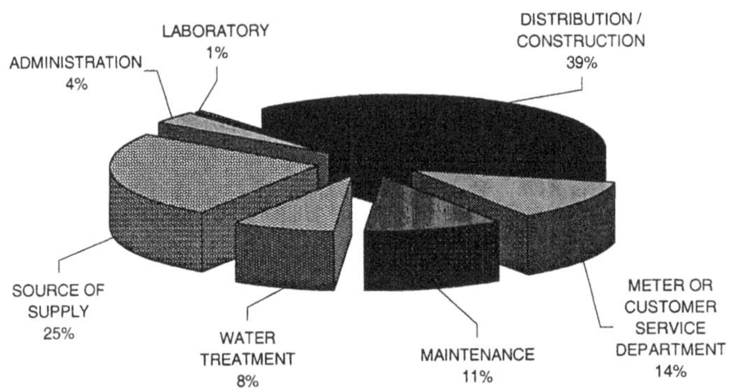

Figure 3.10 Percent Recordable Cases with Lost Workdays for each type of operation (*Source*: Data from AWWA, 1999)

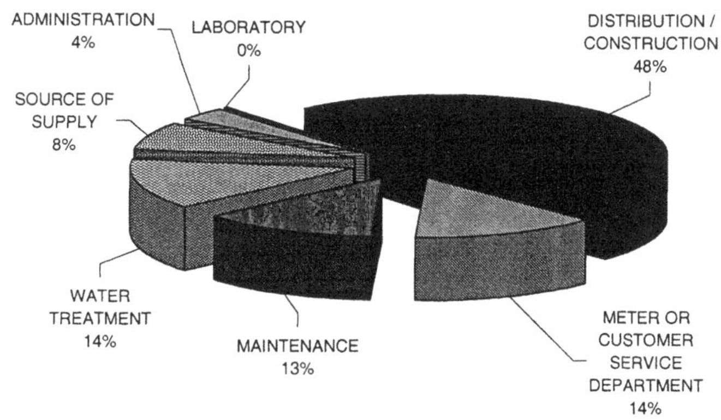

Figure 3.11 Percent of Lost Workdays for each type of operation (*Source*: Data from AWWA, 1999)

F-DE&S Health and Safety Survey

The health and safety survey developed by F-DE&S for this project was designed to provide additional insight into the effects of other factors not explored in prior survey data including health and safety roles and responsibilities, management commitment to safety, and health and safety program attributes.

The purpose of the survey and evaluation was not to make statistically valid comparisons, but to identify emerging health and safety issues in the water industry for further review by the participants in the health and safety workshop.

A copy of complete survey has been included as Appendix B. The survey was divided into seven sections as follows:

- General Information: The purpose of this section was to identify the different types of systems in the water industry and the general approaches used by them for health and safety needs.
- Health and Safety Program Roles and Accountabilities: The purpose of this section was to characterize approaches and commitment to health and safety management.
- Health and Safety Program Attributes: The purpose of this section was to identify the various features of a company's health and safety program.
- Injuries/Illnesses and Lost Work Days in the Water Industry: The purpose of this section was to categorize injuries/illnesses and lost workdays with different types of operations/departments.
- Motor Vehicle Accidents: The purpose of this section was to collect data on utility motor vehicle accidents resulting in death, injury, or property damage.
- Classification of Injuries/Illnesses: The purpose of this section was to identify what the company considers the most hazardous job tasks and why.

- Health and Safety Research and Development Needs: The purpose of this section was to identify areas which the responding company saw a definite need for health and safety research.

Organizations Responding to the Survey

The survey was publicized through mailings and internet postings, but had a limited budget to promote high response. The F-DE&S Survey gathered data from 20 utilities from all around the country. Utilities ranged from water only to combined utilities; from government owned and operated to privately owned and operated utilities; and from large to very small organizations. The following outlines some of the key characteristics of the responding organizations:

- 80% Government owned, and 20% privately owned;
- 30% Subject to Federal OSHA Regulations, 50% exempt from OSHA but subject to state regulations, and the rest believed they were exempt from OSHA and did not have any state regulations;
- Organization size varied, from 10 to 7,000 employees;
- Total of approximately 22,000 employees was represented from within these 20 utilities.

Health and Safety Management Practices

The F-DE&S Survey identifies different types of health and safety management practices that are used in the water industry. Figure 3.12 through Figure 3.18 summarizes key management approach responses obtained from the 20 participating utilities.

Figure 3.12 summarizes methods used to define health and safety performance expectations and accountability. It shows that executive management sets the health and safety policy in about 2/3 of these utilities. About half of the utilities hold employees accountable for meeting measurable health and safety goals. Approximately 40% of all the utilities surveyed hold the employees accountable for non-measurable goals.

Although the executive management sets health and safety policies in most of the surveyed utilities, Figure 3.13 shows that almost all of the executives (95%) delegate responsibilities for health and safety communication to their health and safety staff. More than half of the executives have health and safety on a regular meeting agenda, but less than 1/3 participate in health and safety meetings.

The safety manager was identified as having the overall responsibility for health and safety program oversight and monitoring in 70% of the utilities surveyed (Figure 3.14). More than 2/3 of the utilities identified the health and safety responsible person as a dedicated position with no other responsibilities or as having primary focus on health and safety (Figure 3.15). 60% of the utilities were identified as having a degreed/certified health and safety specialist on staff (Figure 3.16). This shows that majority of the utilities surveyed take a more proactive stand towards health and safety, although the small group of utilities responding may be more aggressive than water utilities as a whole.

In addition to the roles and responsibilities, various attributes of the health and safety management system were also explored in this survey. Figure 3.17 summarizes the attitude of utilities towards formal health and safety management systems. It shows only 25% of all the utilities surveyed have a formal health and safety management system, 40% are planning to develop one, while 25% do not plan to develop a formal health and safety management system. None of the utilities plan to obtain an ISO 14000 certification for their health and safety management system.

Tracking injury/illness data by departments, facilities, job classifications, and job tasks allows organizations to analyze performance and identify which areas require the most attention and utilize their health and safety resources accordingly. Figure 3.18 shows that 85% of the utilities surveyed collect injury/illness data by department and 45% break down the injury/illness data further into job classification.

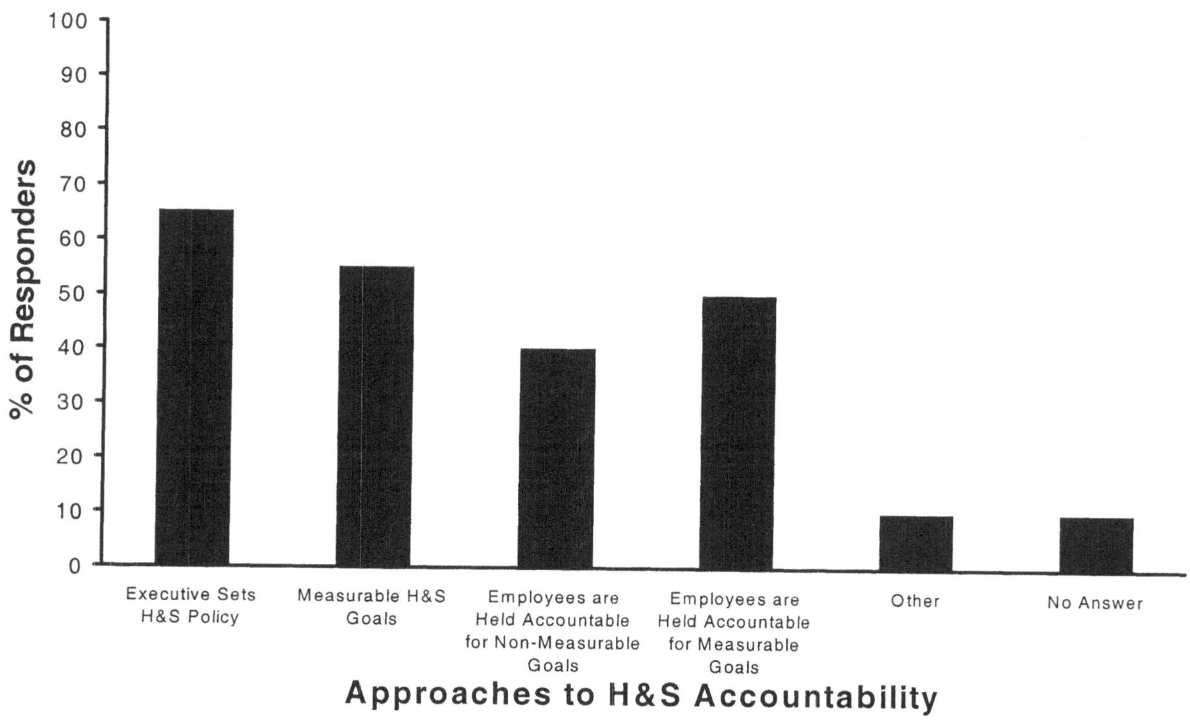

Figure 3.12 Methods used to define H&S performance expectations and determine performance

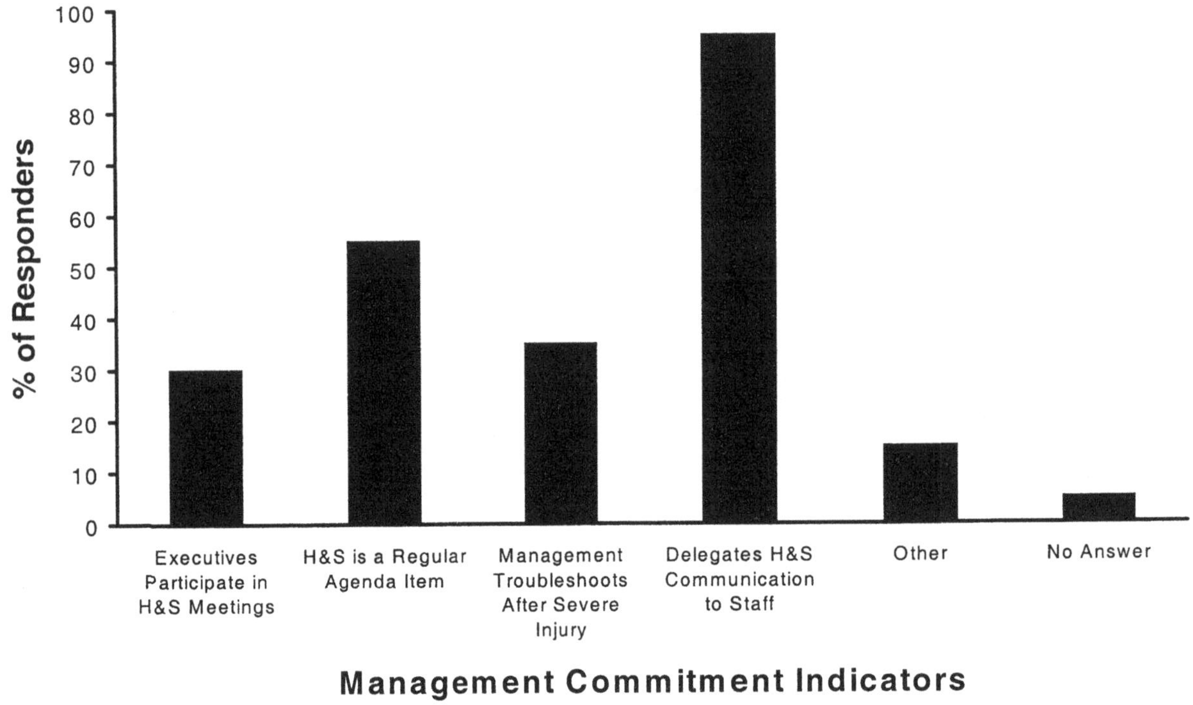

Figure 3.13 Methods used by executive management to communicate interest in H&S

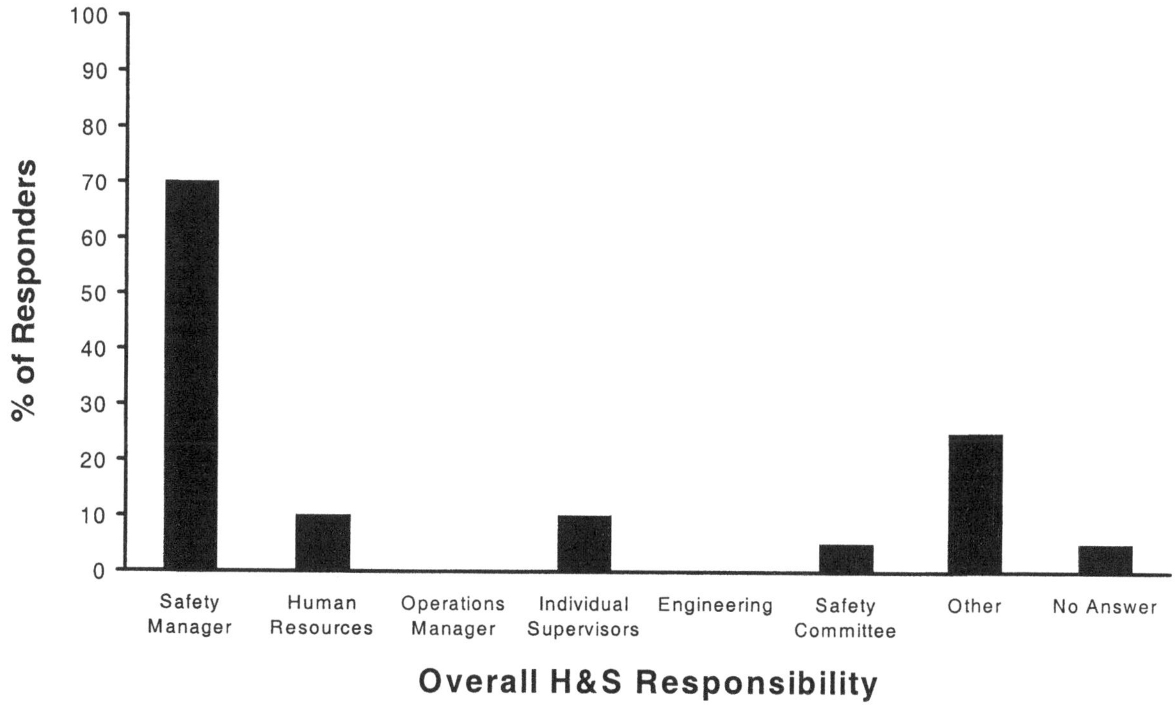

Figure 3.14 Overall responsibility for H&S program oversight & performance monitoring

Figure 3.15 Duties of overall H&S responsible person

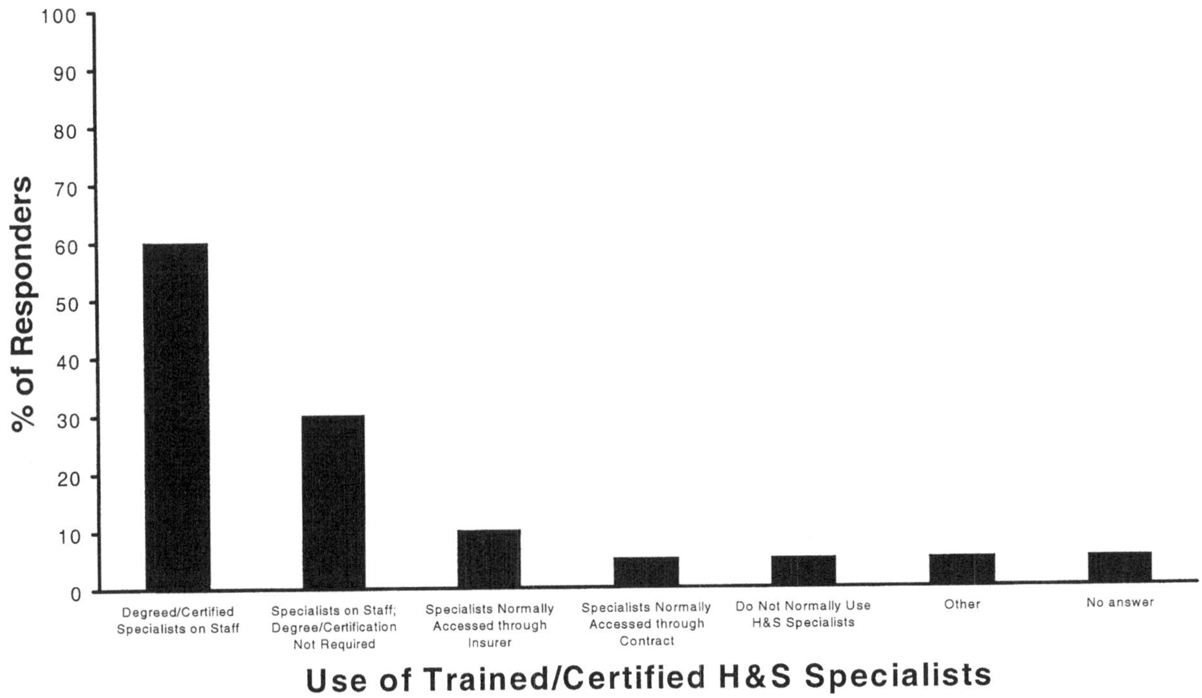

Figure 3.16 Use of trained or certified H&S specialists

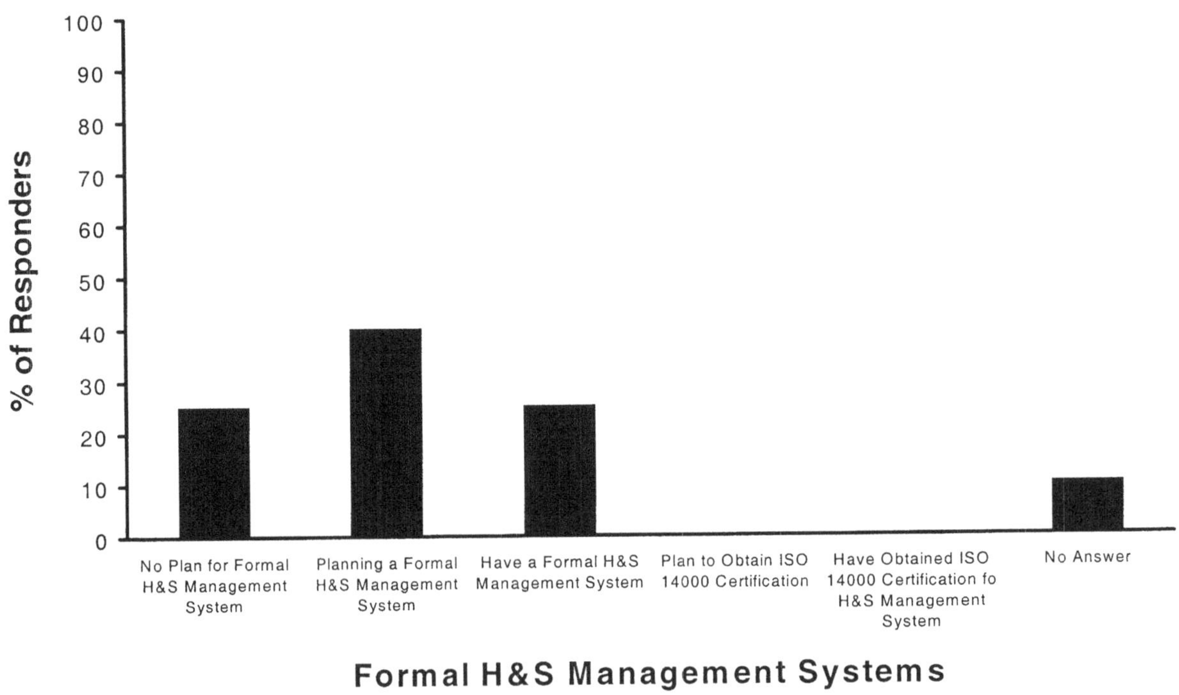

Figure 3.17 Organization's approach to formal H&S (or EH&S) management systems

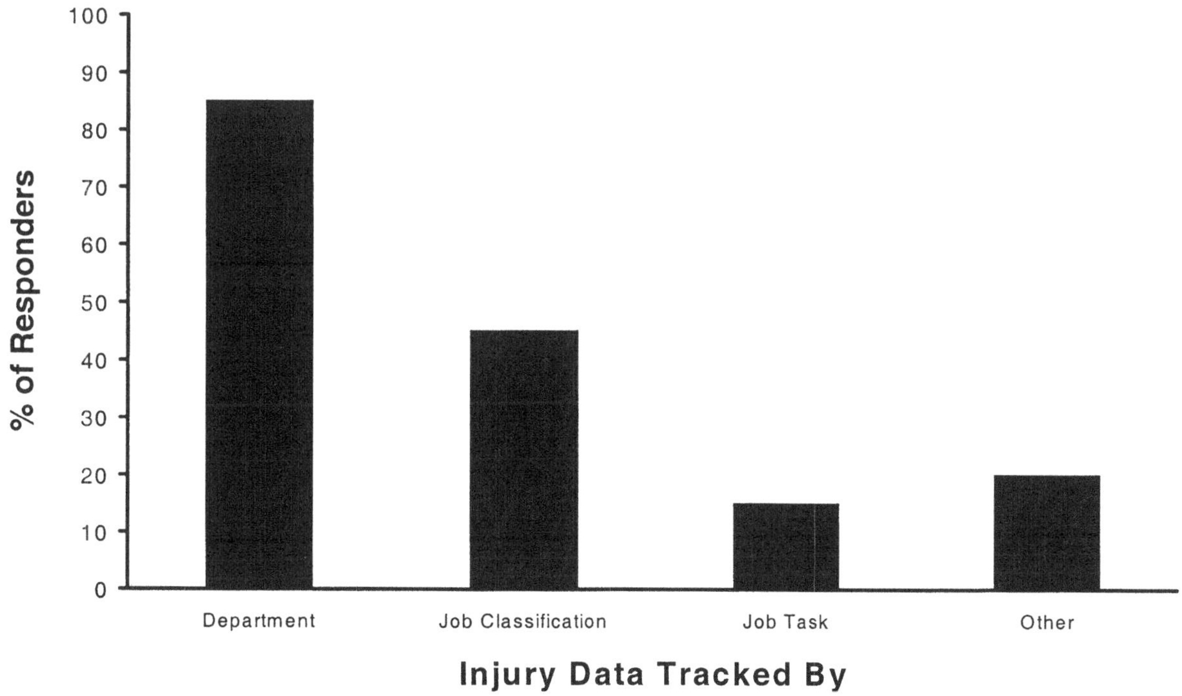

Figure 3.18 Injury/Illness data collection

Where Injuries Occur

Figure 3.19 through Figure 3.21 shows the combined recordable rate, lost workday cases rate, and severity rate, respectively, of the 20 utilities surveyed. The three department/operations with the highest rates in all three of the figures include:

- Distribution-maintenance[3],
- Surface water treatment, and
- Meter[4].

These rates are normalized to represent the approximate number of cases or lost-time days per 200,000 hours worked (i.e. about 100 full-time employees). Hours were estimated based on the reported number of full-time employees and an estimated 1,840 hours worked per year for a typical employee. This provides a more balanced assessment of performance for smaller departments than

[3] Distribution-maintenance includes maintenance of pipelines, hydrants, etc. in distribution systems (not meter).

[4] Meter includes actually reading meters (not billing department), meter installation/repair etc.

in Figure 3.9 through Figure 3.11, which only consider the total number of cases or lost-time days. Figure 3.19 through Figure 3.21 suggest that meter and surface water treatment have similar needs for performance improvement as distribution-maintenance.

Each of the 20 utilities were asked to list two departments, and five job classifications that they considered to involve the most hazardous work. The responses shown in Figure 3.22 and Figure 3.23 suggests that responders' perceptions are generally consistent with the recordable rate, lost workday cases rate, and the severity rate with the exception of meter reading. Meter reading was much less often identified as a "most hazardous" department, perhaps because of the smaller size of this department. The data showed that incidence and severity rates are high enough to rank it relatively evenly among the top three departments with the greatest health and safety needs.

Figure 3.24 summarizes what the responders considered to be the most hazardous job tasks. The following are among the top four in the list:

- Working in excavations or excavating
- Material handling/lifting
- Confined space entry
- Driving vehicles

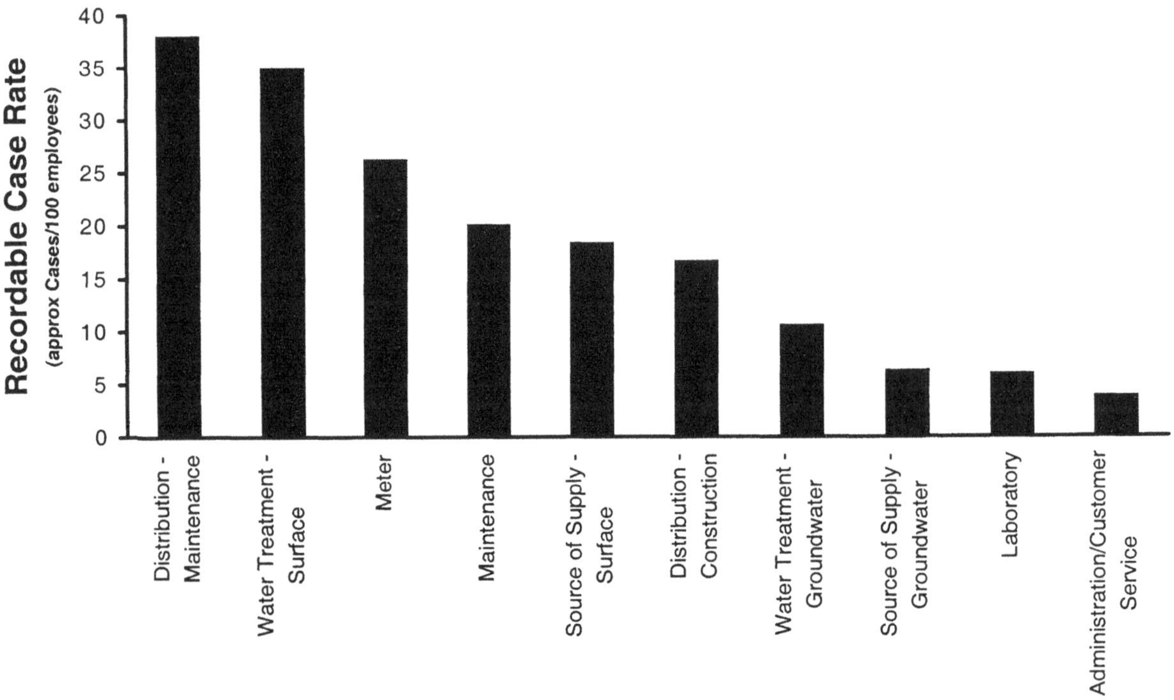

Figure 3.19 Recordable Rate

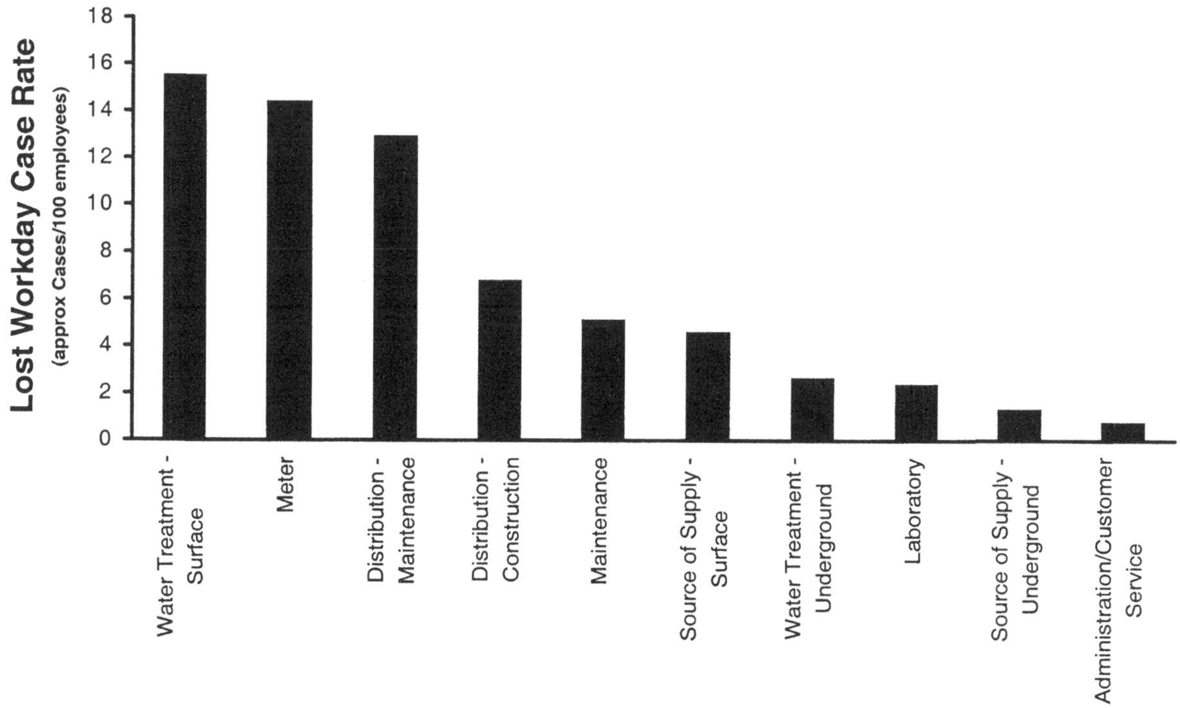

Figure 3.20 Lost Workday Case Rate

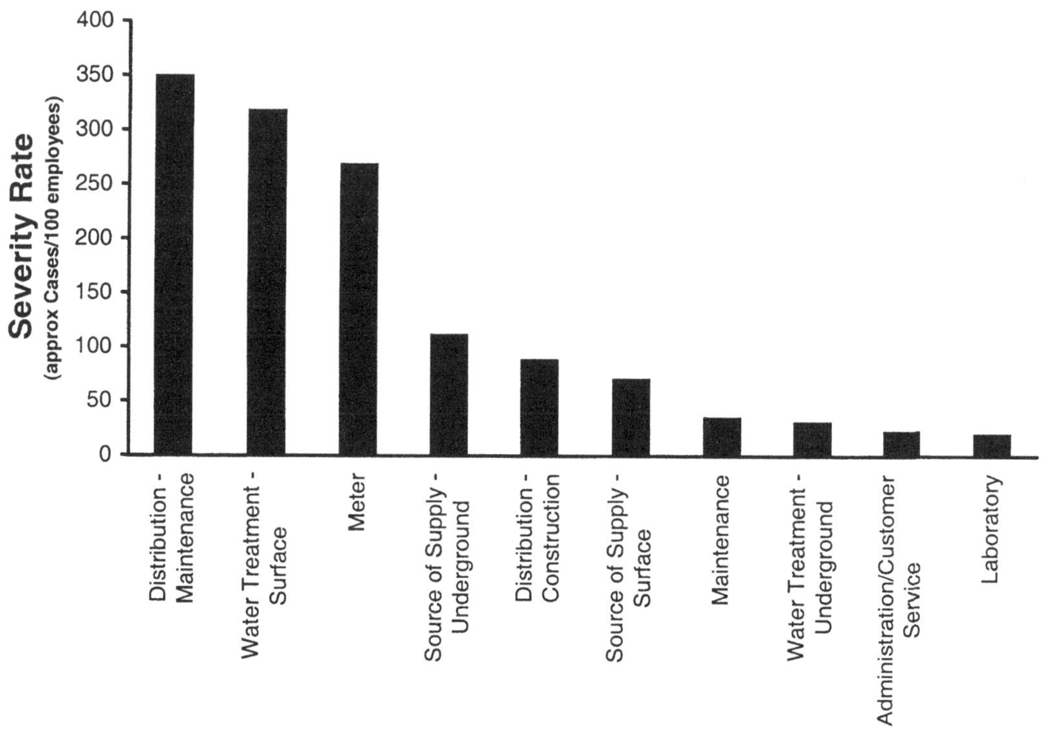

Figure 3.21 Severity Rate

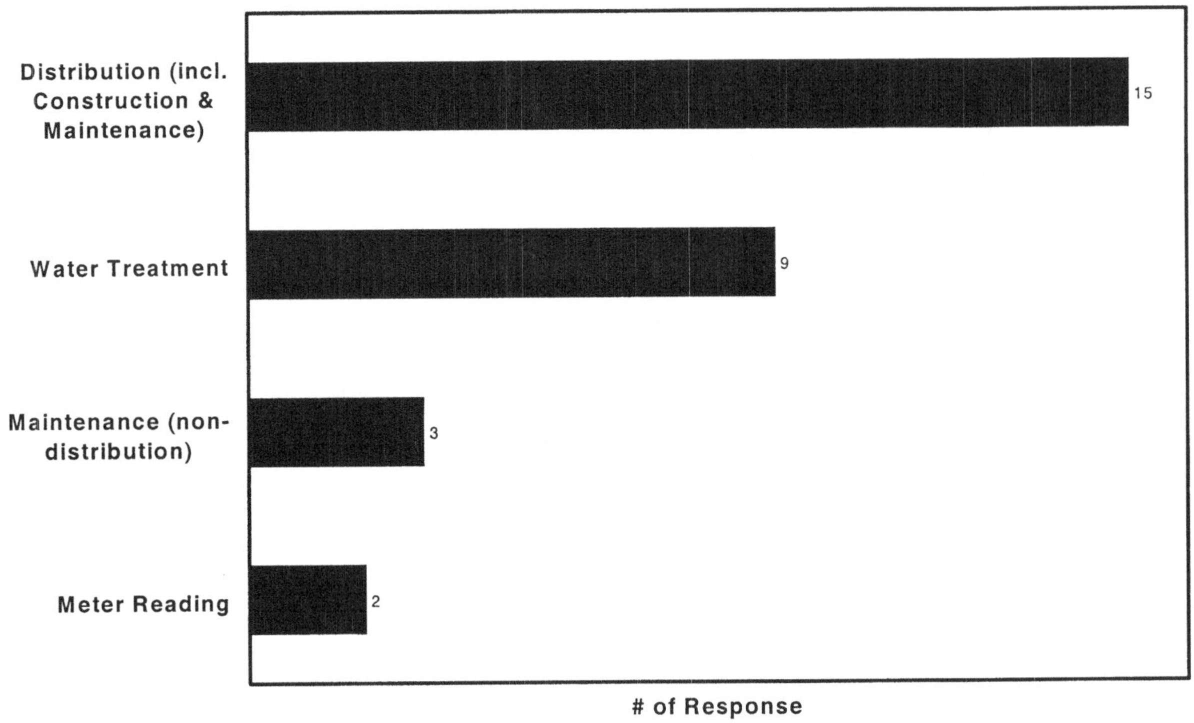

Figure 3.22 Perceptions of most hazardous departments

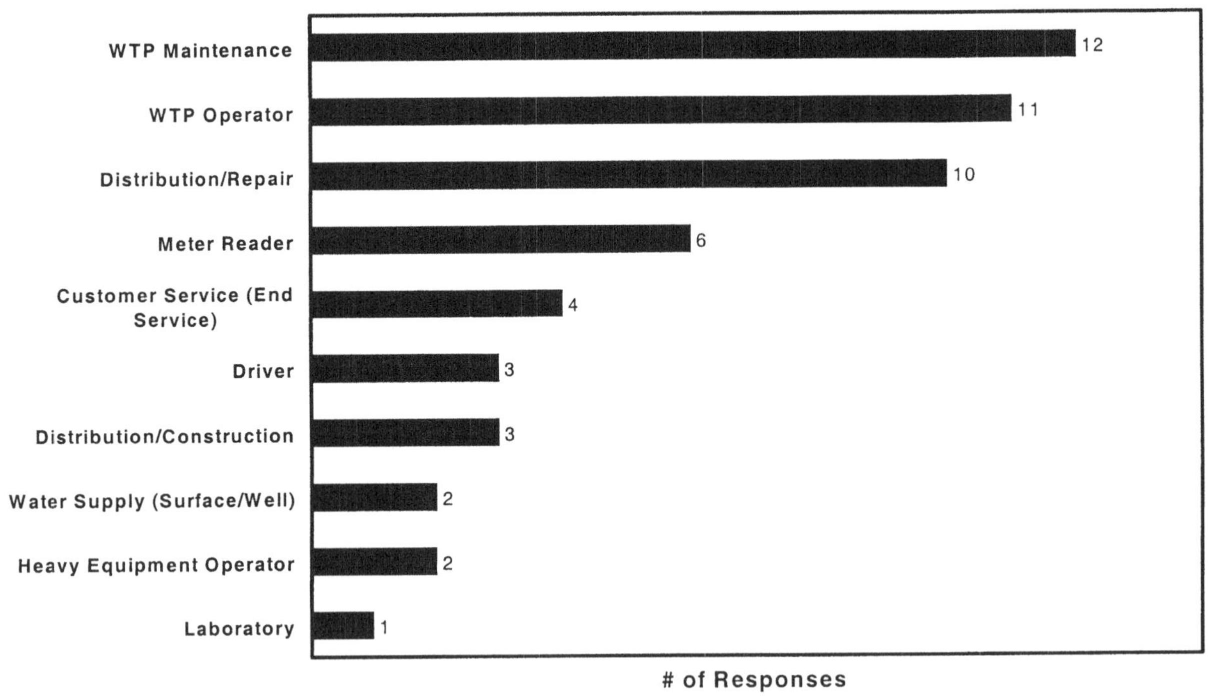

Figure 3.23 Perceptions of most hazardous "job class"

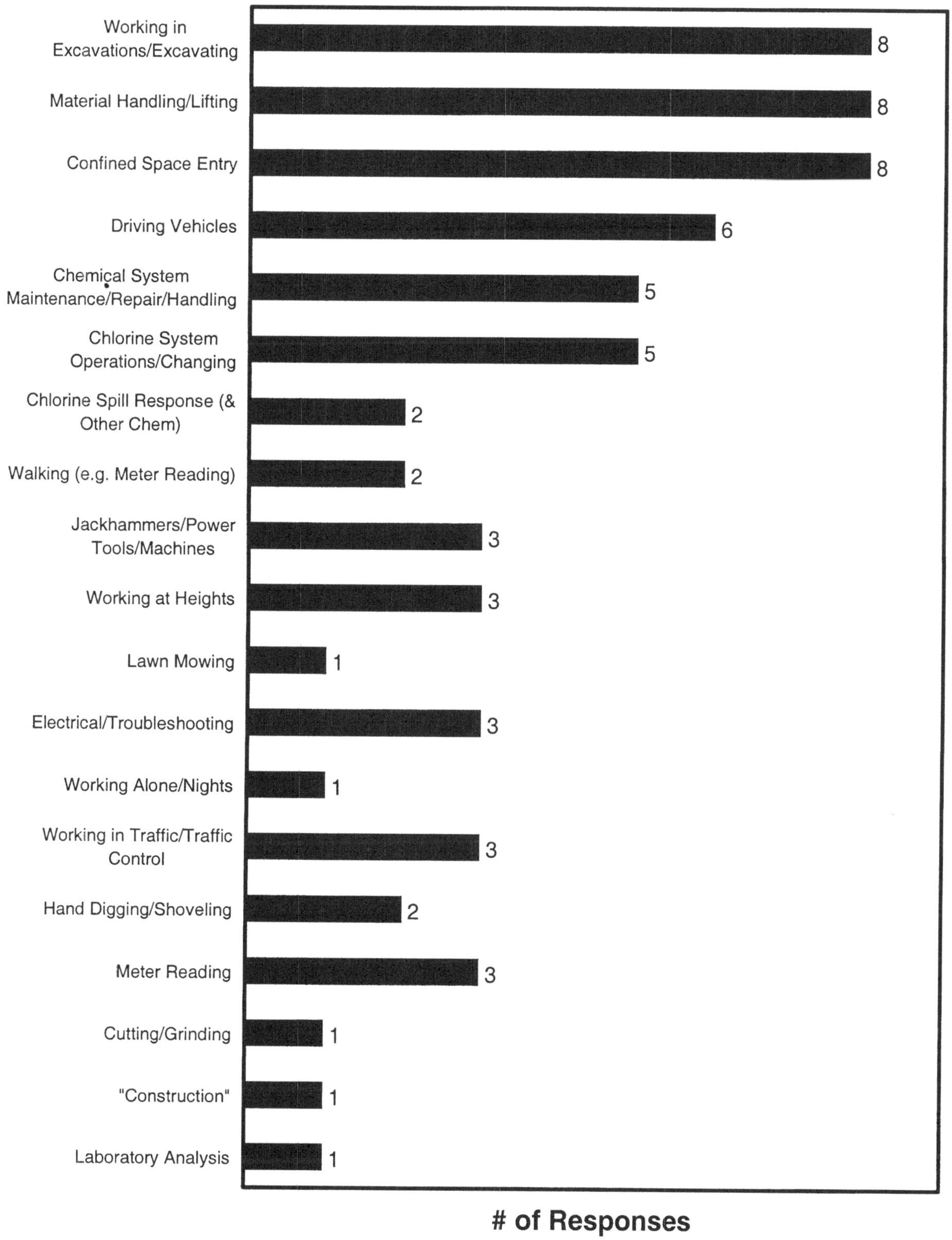

Figure 3.24 Perceptions of most hazardous job tasks

Greatest Health and Safety Needs

Utilities were asked to identify the top health and safety needs of their organizations. Table 3.2 below presents areas that require the greatest health and safety needs as identified by the 20 utilities surveyed. The top three health and safety needs are for training, management commitment, and formal written health and safety programs.

Table 3.2
Most Important Health & Safety Needs

Category	Rank	# of Times Cited
Training	1	14
Management Commitment/Involvement	3	13
Formal, Written H&S Programs	3	13
Accountability	5	8
Safety Culture/Employee Involvement	5	8
Sufficient H&S Staff/Resources	6	7
Specific H&S Problems (Technological & Administrative)	7	6
Safety Incentive Programs	8	4
Audit/Assessment and Corrective Action Programs	10	3
Measuring Performance	10	3
Strong Incident Reporting, Evaluation and Corrective Action	11	2
OSHA Compliance	12	1
Injury Case Management and Rehabilitation Programs	13	0

CONCLUSIONS OF THE EVALUATION

Readers are urged to be cautious in the use of the data presented and hypotheses developed because of the limitations of the AWWA Safety Survey and the limited sample size of the survey conducted by F-DE&S for this project. The comparisons made and hypotheses developed primarily intended to inform the workshop; however, despite the limitations, it is hoped that the information can provide those involved in water utility health and safety with insights that could help reduce the frequency and severity of injuries and illnesses.

If resources are to be allocated based upon the incidence and severity rates (instead of the numbers of incidences as presented in Section 0), water utilities should prioritize their effort and resources as follows:

Highest Priority:

- Distribution Maintenance
- Surface Water Treatment
- Meter

Second Priority:

- Source of Supply
- Distribution Construction
- Maintenance

Third Priority:

- Ground Water Treatment
- Administration
- Laboratory

CHAPTER 4
WORKSHOP PRESENTATIONS AND RESULTS

A two-day health and safety workshop was held for water utilities and other stakeholders to help identify emerging health and safety issues, identify future health and safety research needs and develop a research and development agenda for the water industry.

The workshop was organized to provide a mixture of presentation, discussion and structured brainstorming to accomplish the project objectives. The workshop generally included:

- Presentations on existing injury/illness data for the industry.
- Presentations on the results of the survey conducted for this project by F-DE&S.
- Presentations on relevant work by others in the field of health and safety research, safety management and excellence programs, and on other critical safety issues for the water industry.
- Brainstorming to identify potential solutions to address those critical issues that can be developed through research.
- Ranking the priority of potential research projects.
- Developing several project descriptions.

The following sections describe the proceedings of the workshop.

WORKSHOP PARTICIPANTS

To ensure that the results of the workshop are broadly meaningful, useful and will have positive impacts for the water industry, a variety of stakeholders in water industry safety and injury prevention were invited to participate in the workshop. Organizations participating, detailed in the Acknowledgements section, included:

- 20 water utilities that responded to the F-DE&S Survey. These utilities, as mentioned earlier, represented small and large, public and private, and geographically diverse utilities with a broad range of health and safety approaches and resources.
- Government agencies that both enforce occupational health and safety requirements and provide support and expertise.
- Critical organizations that support accident prevention and through their body of knowledge, capabilities and expertise. This included relevant professional and industry associations.
- Representatives from a power utility that presented on exemplary practices in the power industry and a power industry safety incidence information sharing consortium.

INSIGHTS FROM TECHNICAL PRESENTATIONS

The workshop included presentations that covered both the water industry injury/illness data evaluation and project survey results. These presentations highlighted key issues that were

identified by the F-DE&S survey respondents and the workshop participants as critical or emerging issues. Speakers also provided input on their own experience and expertise with injury/illness statistics as well as insights to water utility execution and H&S staff.
Table 4.1 below lists the workshop presentations.

Table 4.1
Workshop Presentations

Organization	Title
F-DE&S	Introductions and Objectives; Overview of the Brainstorming and Decision Analysis Process
National Safety Council/National Occupational Research Agenda	Traumatic Occupational Injury Research Needs and Priorities
F-DE&S/BLS	Review of BLS Private Water Industry Injury / Illness Data
F-DE&S	Evaluation of Current Water Industry Injury/Illness Data
New York State Department of Public Employee Safety and Health	Water Supply – Safety & Health: Regulatory Perspective
Duke Energy Corporation	Best Practices for Power Industry Data Collection / Analysis
National Safety Council	Key Elements of An Effective Safety Program
San Antonio Water System	Behavior-Based Employee-Driven Safety Program
Florida Workers' Compensation Division	System Outcome Measures for Workplace Disability Prevention

Selected excerpts of presentations that can provide insights on safety needs and how to improve safety performance in water utilities are summarized in the following subsections.

BLS Data on Private Water Company Health and Safety Performance

The Department of Labor's Bureau of Labor Statistics (BLS) collects injury and illness data from a scientifically selected sample of private employers. For example, in 1996, BLS collected and statistically analyzed data from approximately 170,000 establishments. To support the workshop, BLS provided AwwaRF with water industry data for five years, including 1994 to 1996 and 1998 to 1999, and general industry data for 1999.

Figure 4.1 presents the incidence rates for nonfatal cases with days away from work for the private water industry in the five years provided, as well as the 1999 general industry rate. This figure indicates that the incidence rates for the private water industry has steadily declined during these years, but it remained approximately 40% higher than the rate for industry as a whole in 1999.

Percentages of cases with days away from work in the private water industry attributed to each type of employee (using BLS categories for 1996) are presented in Figure 4.2. The BLS categories do not appear to be as meaningful as the functional groups used in the AWWA and F-DE&S surveys, however (e.g., see Figure 3.9 and Figure 3.19).

Non-Applicability of OSHA to Government Employees

The Federal Occupational Safety and Health (OSH) Act of 1970 did not grant the Department of Labor's Occupational Safety and Health Administration (OSHA) authority over Federal, Local or State government-operated facilities. In developing its program to authorize states to enforce Federal regulations, OSHA recognized the need to encourage states to regulate State and Local government workplaces. The process of getting state authority involves submitting an acceptable "State Plan" to OSHA and OSHA has required States to enact legislation and regulations that not only are equivalent to Federal rules, but that also apply to the State and Local Government workplace.

As a result, many states have their own OSHA rules that apply to state and local government-operated WTPs, in particular all those states with OSHA plans approved by federal OSHA (i.e., "OSHA-Plan" states). In addition, some states have enacted equivalent rules that apply only to State and Local Government employees.

The Federal Government's employees are covered under Section 19 of the OSH Act. This has been augmented by Executive Order 12196, dated February 26, 1980, which requires agency heads to furnish employees a workplace free from recognized hazards that are causing or are likely to cause death or serious physical harm.

As a result, Federal or State equivalent OSHA rules are applicable and enforceable if you are either:

- A privately operated plant.
- Operated by the federal government.
- In one of the following states operating OSHA Plans.

Table 4.2
OSHA-Plan States

Alaska	Indiana	Minnesota	Oregon	Vermont
Arizona	Iowa	Nevada	Puerto Rico	Virgin Islands
California	Kentucky	New Mexico	South Carolina	Virginia
Connecticut*	Maryland	New York*	Tennessee	Washington
Hawaii	Michigan	North Carolina	Utah	Wyoming

* OSHA Plan State with enforcement authority for public sector only.

Note that some states (e.g., Ohio) have enacted OSHA equivalent rule that apply to the public sector, but do not have State Plan approval.

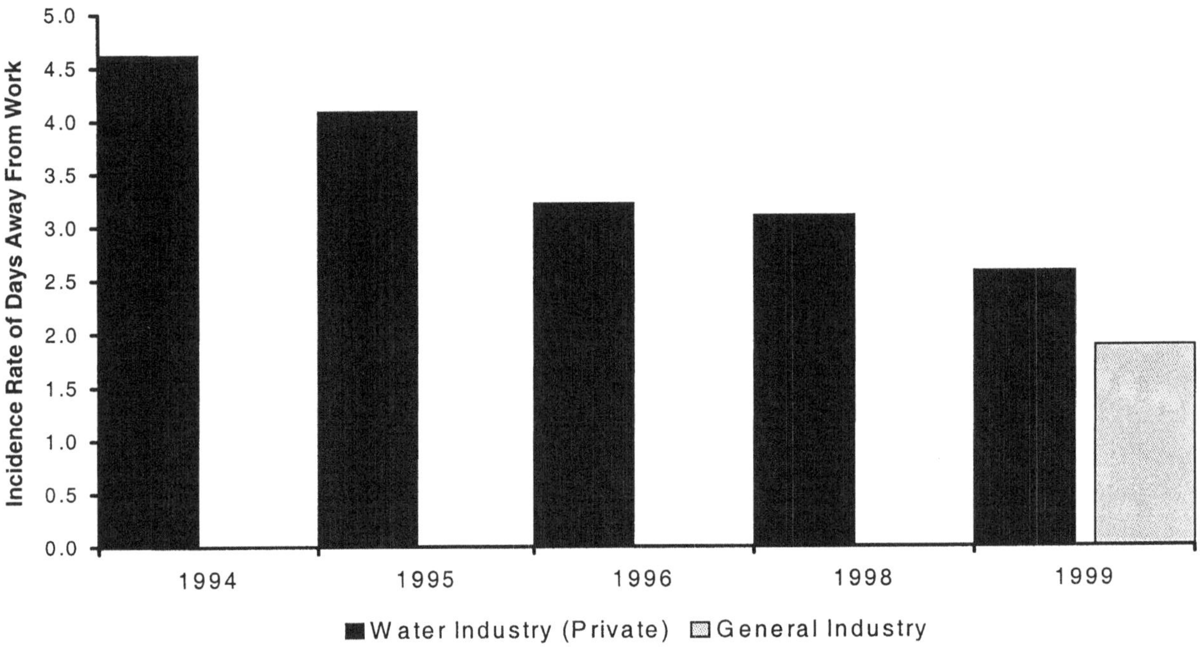

Figure 4.1 Incidence Rates of nonfatal cases with days away from work for private water industry (1994-1999) vs. general industry (1999) (*Source*: BLS, 2000)[5]

[5] Note: 1997 data is not available.

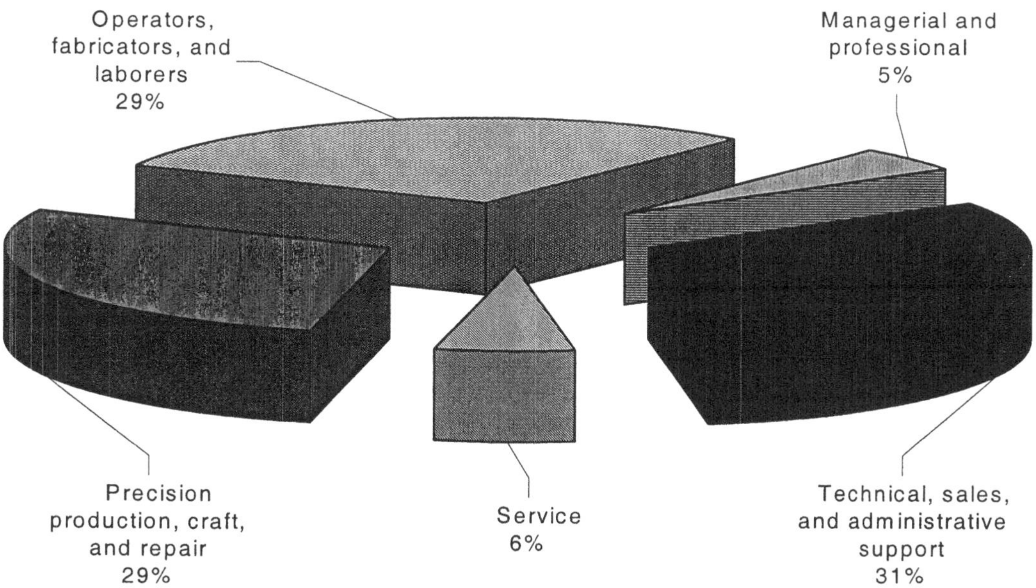

Figure 4.2 Distribution of nonfatal cases with days away from work by type of employee for private water industry in 1996 (*Source*: BLS, 2000)

The Department of Labor's Office of Inspector General published a report in February 2000 entitled *Evaluating the Status of Occupational Safety and Health Coverage of State and Local Government Workers in Federal OSHA States.* (Report Number 05-00-0001-10-001). This report concluded that:

- Many of the 29 States and territories that do not have an approved State Plan lack legislation, regulations, enforcement authority, adequate staffing or other critical elements of Safety and Health coverage.
- There is limited data in these States and territories on the occurrence of injuries and illnesses among public sector employees with which to evaluate Safety and Health programs.
- Data from State Plan States suggest that Local Government workers are experiencing significantly higher occupational injury and illness rates compared to State workers

The report recommended that OSHA do more to encourage and promote state programs in non-Plan States. The Department of Labor Bureau of Labor Statistics has embarked on additional data collection and evaluation of this problem, but has indicated that it has little authority to effect change under current law. Initial evaluation results reportedly confirm higher injury/illness rates among government employees and higher rates in non-Plan States; however, these results have not been released because of the difficulty in confirming their statistical confidence intervals.

A bill before Congress, H.R. 776, would extend Federal OSHA coverage to employees of States and their political subdivisions.

A Public Entity Regulator's Perspective

One perspective shared at the Workshop was of a State regulator from an agency focused exclusively on public sector employees, the New York State Department of Public Employees Safety and Health (PESH). While based on personal experience and not State policy or regulation, the PESH participant shared interesting perspectives on some potential underlying factors that may pertain to public sector work environment that may be different from the private sector. These include:

Cultural Factors

- We are the Government & a monopoly & exempt!
- We do not have to be elected
- We perform a vital function
- It has to be done NOW

Employee Attitudes

- We have always done it that way
- I just work here
- You can't fire me

Institutional Factors

- Cost of prevention is a budget item
- Cost of injury is someone else's budget item
- No accounting for injury costs

Research Needs

- What are the costs of injury and illness?
- Who pays for what?
- How do you change the culture?
- What role should regulation play?

These observations were generally considered valid by the Workshop attendees, including public utility Health and Safety representatives, although the degree to which these factors affect performance varies widely at public utilities, depending upon its culture and management emphasis on safety programs and improvement.

Best Practices in the Electric Utility Sector

Another perspective shared at the Workshop was from a power utility that is applying best practices for collection and analysis of injury/illness data. This utility had participated in an initiative of the Electric Power Research Institute (EPRI) to develop an injury/illness data-sharing consortium. These best practices shared at the Workshop include:

Objectives of Injury / Illness Data Collection

- Monitor trends over time
- Identify occupations with higher injury /illness rates
- Quantify costs in terms of dollars, lost workdays
- Evaluate effectiveness of injury / illness programs
- Determine research priorities
- Benchmark performance with other companies, associations

Computerized Safety Information Management System

- "All-type" incident reporting
- Real time
- Web based
- Incidence analysis, trending and graphing capability
- Reporting - Regulatory and Informational

State-of-the-Art Analysis

- Traditional data - Injury classification, injury nature, body part, costs, lost workdays
- Second tier data - job class codes, hours worked
- Separate into specific rates per job class, type of injury, and demographic groups - not company.
- Rates can then be compared to workforce characteristics to determine which populations are most at-risk.
- Observations by specific categories (job type, gender, age group, or time) can identify opportunities for targeted intervention and cost reduction.

EPRI Health and Safety Database

- Begin to address research needs of electric power industry
- Integrates personnel, injury and claims data into one data system
- Summarizing injury trends by company, occupation, injury type, and demographic factors (age, gender).
- Initially integrated 5 years of data (1995-1999)
- Now contains 135,000 employee-years of data

Benefits of Health and Safety Data Consortium for Participating Utilities

- Begin with existing data
- Integrate new data as it becomes available.
- Consistent format requires no special work-up of data
- Report at any convenient time
- Determine effectiveness of specific H&S programs on reducing risk, H&S costs or lost work.

- Detailed analysis for specific injury types (lower back injury, repetitive motion injury) to target intervention
- Derive specific injury/illness characteristics for selected occupations or work locations
- Benchmarking your organization against similar utilities

Examples of how this kind of data sharing can be used to benchmark an organization's performance against peer utility groups are provided in Figure 4.3 and Figure 4.4.

National Safety Council on Key Elements of Successful Safety Programs

The Chair of the Utilities Division of the National Safety Council shared NSC's views of the key elements of successful safety programs. Key perspectives shared include:

Critical Success Factors

- Written policies, programs and procedures
- Communicating a shared vision of safety as a value
- Expect participation & results from all levels
- Expressing Management Commitment through both involvement and allocation of resources
- Accountability of management and employees through enforcement of policies and programs
- Tracking and evaluating performance and setting goals for improvement

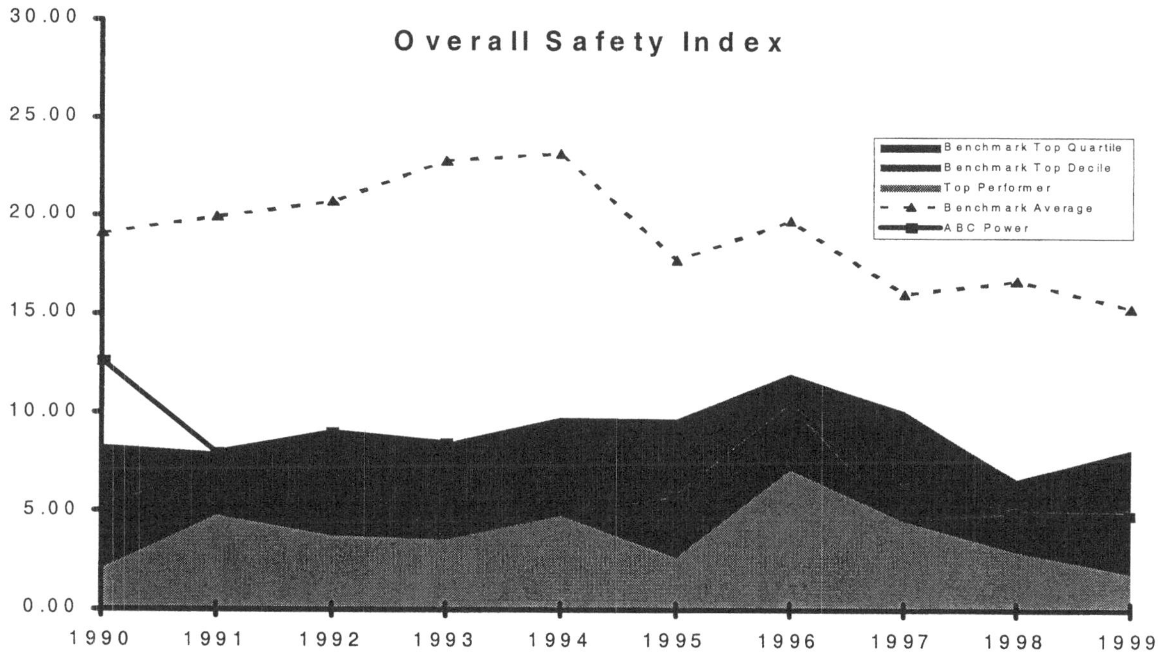

Figure 4.3 Benchmarking using organization-wide H&S performance indicator (*Source*: Almond, 2001)

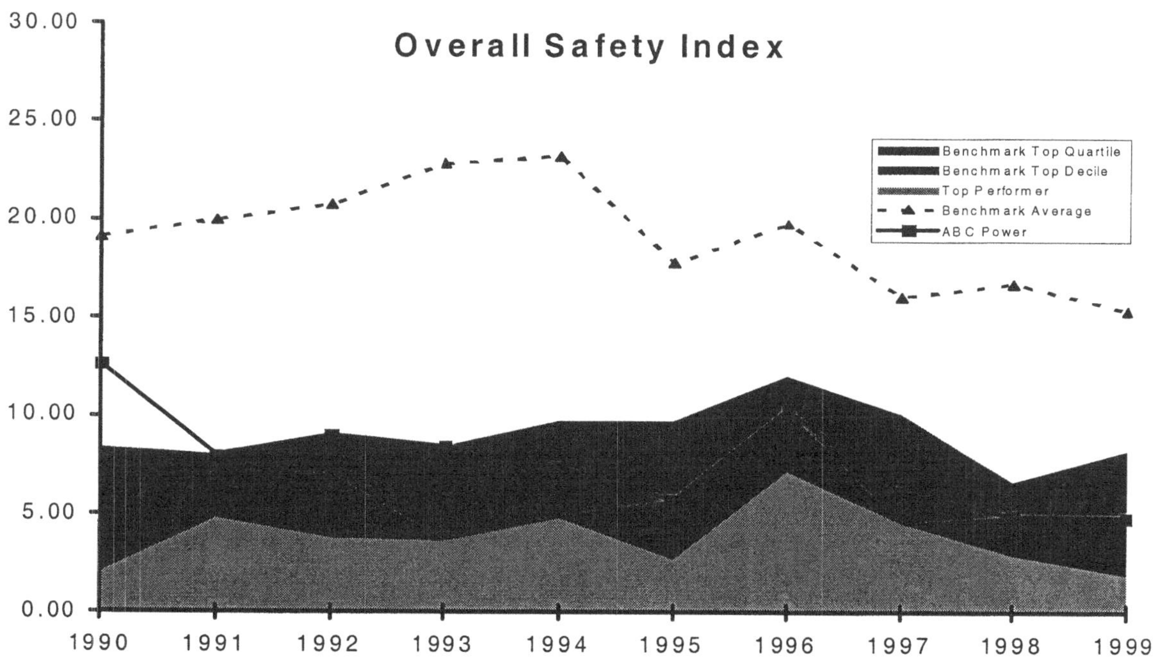

Figure 4.4 Benchmarking a specific operating unit/specific performance indicator (*Source:* Almond, 2001)

The National Safety Council provides detailed guidance on safety program excellence in the NSC manual *14 Elements of a Successful Safety and Health Program.* These elements include:

NSC 14 Elements of a Successful Safety and Health Program:

1. Hazard Recognition, Evaluation and Control
2. Workpace Design and Engineering
3. Safety Performance Management
4. Regulatory Compliance Management
5. Occupational Health
6. Information Collection
7. Employee Involvement
8. Motivation, Behavior and Attitudes
9. Training and Orientation
10. Organizational Communications
11. Management and Control of External Exposures
12. Environmental Management
13. Workforce Planning and Staffing
14. Assessments, Audits and Evaluations

NSC also presents a continuous improvement model that includes:

NSC Continuous Improvement Model:

Phase 1 – Management Commitment and Involvement
Phase 2 – Establish a Baseline
Phase 3 – Set Goals
Phase 4 – Implement Strategies
Phase 5 – Review and Adjust

San Antonio Water System on Employee-Driven Behavioral Safety

San Antonio Water System (SAWS) presented on the methods used and results from its successful behavior-based, employee-driven safety improvement program initiated in 1997. SAWS felt that past programs did not have employee "buy-in" because they were top down programs without significant employee involvement. SAWS made a commitment to a behavior based program run by employees that provided "bottom-up" safety recommendations to management. Also critical was management commitment and support, willingness to change behavior and a recognition and award program.

A Safety Committee was created at each facility with 5-8 non-management members representing each type of operation conducted. The committee was led by a non-management member and members also served as behavior "observers". The committee functions included:

Facility Safety Committee Functions:

Obtain Data:
- Review Accident Reports
- Review Observation Reports
- Review Safety Suggestions

Analyze Data:
- Evaluate Data Quality
- Determine Critical Behavior
- Determine Training Needs
- Determine Equipment Needs
- Determine Process Changes

Provide Feed Back:
- Prepare Written Reports
- Critique Observer's Reports
- Critique Accidents
- Provide Recommendations
- Forms Modification
- Equipment Modification(s)
- Process Modification(s)
- Facilities Modification(s)
- Safety & Job Skill Training

Management's Role:

Provide Resources:
- Personal Commitment
- Budget Related Items

Insure Objectivity:
- Follow Program Guidelines
- Look At Trends
- Assure Confidentially of Data

Make Decisions:
- Timely Decisions at all Levels
- Support Valid Reasons

Follow-Up:
- Assure Decision Implemented
- Analyze Results

- Written Implementation (Who, What and By When)

Provide Communication Linkage:

- To Upper Management
- To Lower Management
- To Facility's Safety Committee

Results in terms of reduction of lost time injuries and workers compensation costs were dramatic. Results are summarized below.

San Antonio Water System Improvements Through Behavioral Safety Program:

Year	Number of Lost Time Incidents	% Reduction in Workers Comp. Cost
1997	70	--
1998	25	50.1%
1999	23	68.2%
2000	19	40.3%

(*Source*: Data from Skidmore, 2001)

Florida Division of Workers' Compensation on Performance Measures

The Florida Division of Workers Compensation, Bureau of Research and Education presented on measuring factors that affect the outcomes of workers compensation cases. While the other presentations were focused primarily on prevention of injuries and illnesses, this presentation also illustrated the importance of good case management and rehabilitation in minimizing long-term disabilities and their associated costs. Post-incident management does not receive sufficient attention in many organizations, and the cost of any incidents can be increased by inadequate management in the post-incident phases of disability prevention.

Figure 4.5 illustrates the critical aspects of disability prevention and management that go well beyond the prevention stage. Figure 4.6 through Figure 4.8 illustrate the differences in selected performance measures between "good" and "bad" workers compensation insurance carriers in Florida in the 1990s.

IDENTIFICATION AND PRIORITIZATION OF RESEARCH PROJECTS

Following the workshop presentations, a professional facilitator guided the workshop attendees through a structured brainstorming process to identify critical health and safety issues in the water industry, and their potential solutions. The workshop participants prioritized these issues/solutions and developed work scope descriptions for the highest-ranking potential research projects. The following subsections describe these steps in more detail.

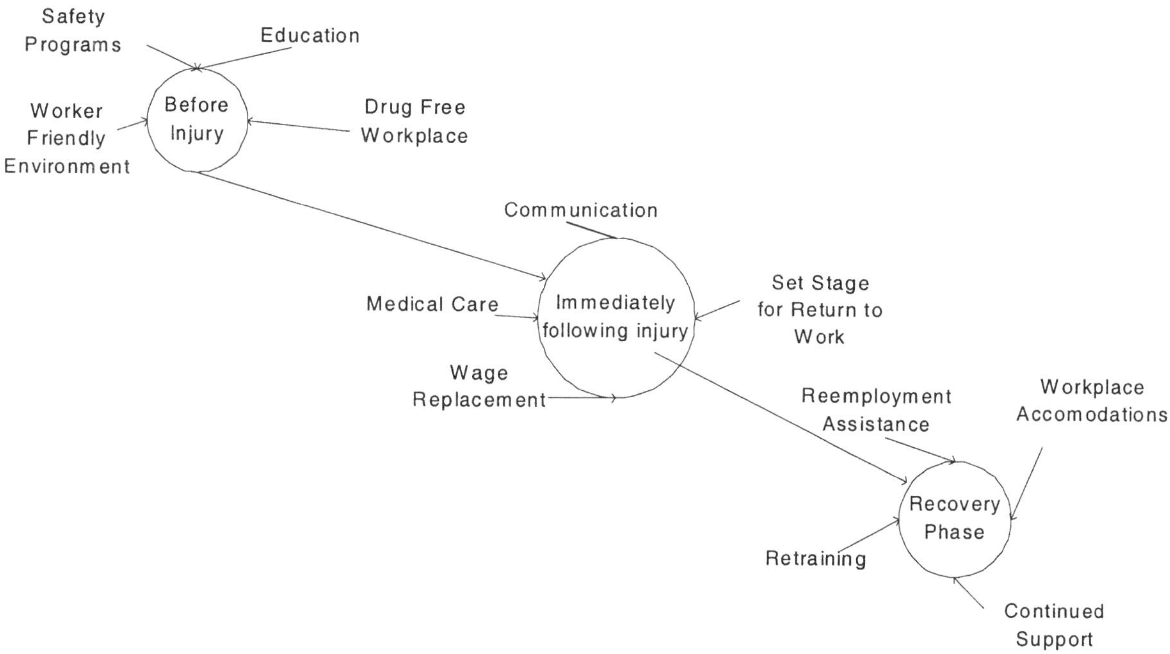

Figure 4.5 Critical aspects of disability prevention and management (*Source*: Markiewicz, 2001)

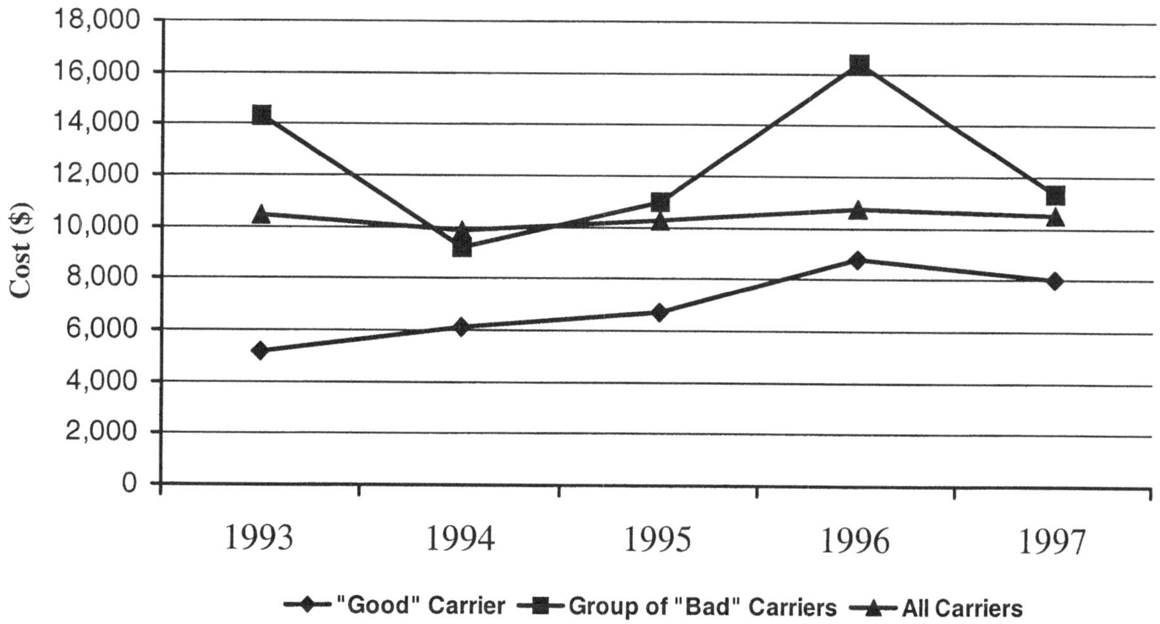

Figure 4.6 Average total cost per Lost Time accident after 18 months (*Source*: Markiewicz, 2001)

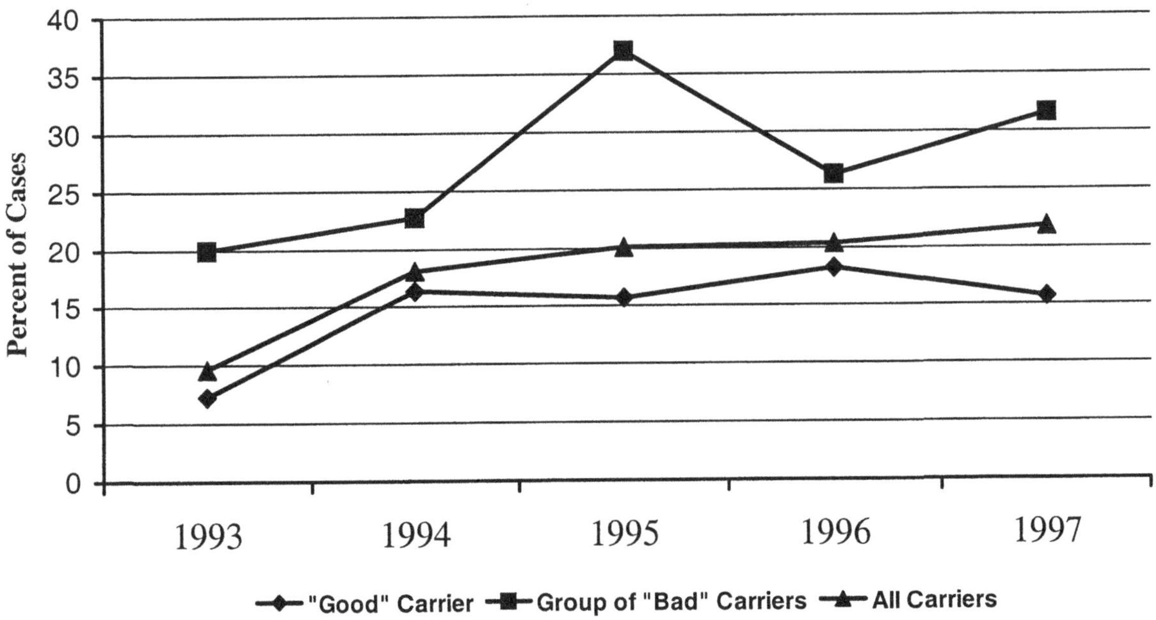

Figure 4.7 Percent of cases classified as permanent partial disability after 18 months (*Source*: Markiewicz, 2001)

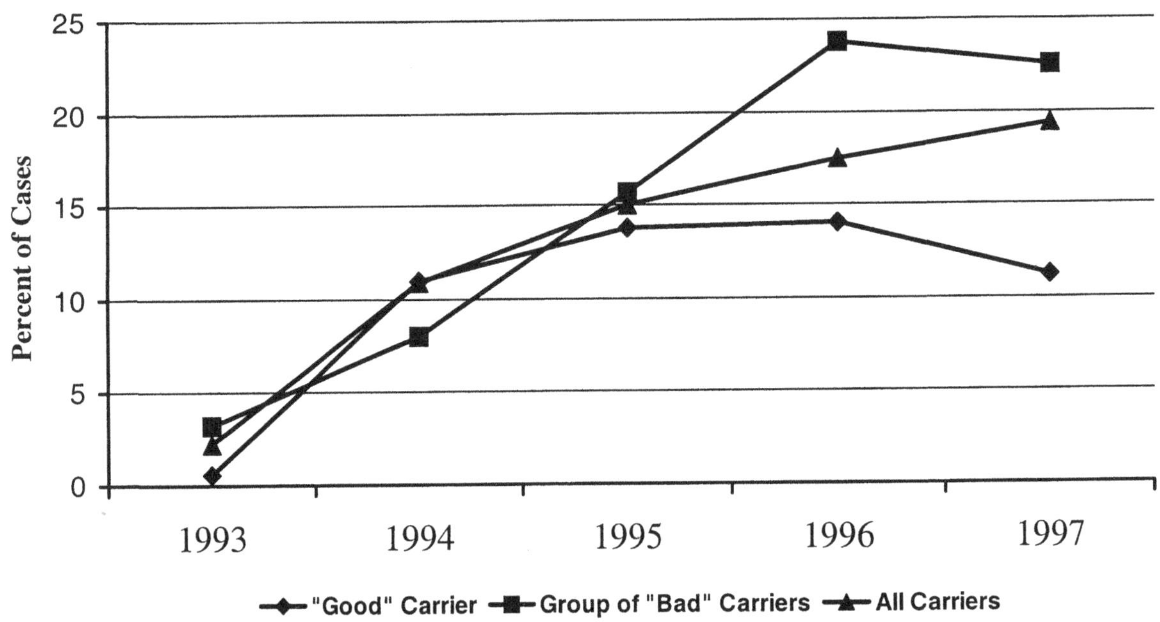

Figure 4.8 Litigation Rate after 18 months (*Source*: Markiewicz, 2001)

Identification of Potential Research Projects

Workshop participants were informed in the identification of critical issues and potential solutions by the data evaluations discussed in Section 3 and the presentations discussed earlier in Section 4. Limitations in the available sources of data and the benefits of the power industry data consortium led to the identification of improved data collection in the water industry as a priority.

The breakdown of injury data by type of work and type of injury (see Figure 3.22 through Figure 3.24) suggested emphasis on traumatic back and shoulder injuries. The NIOSH National Occupational Research Agenda and the list of most important health and safety needs derived from the survey (Table 3.2) resulted in workshop participants identifying several research project areas, including "emerging issues" faced by the water industry (such as security, terrorism, stress, culture changes, work organization structure, ergonomics, etc.).

With the help of the facilitator, the participants organized the results of the brainstorming, developed a prioritization system and then narrowed the potential research projects down to those highest priority projects for the 5 to 10 year horizon. The eight highest priority projects and the purpose of these projects are presented in Table 4.3.

Table 4.3 Potential Research Projects Identified

Potential Research Project	Purpose of the Project
1. Effectiveness of Techniques for Prevention of Lower Back and Shoulder Injuries and Other Musculoskeletal Disorders	To identify the most effective techniques (and possible improvements to existing techniques) to prevent lower back, shoulder and other MSDs in the water industry (e.g. evaluating the use of backbelts, field hoists on trucks, etc.)
2. Best Practices for Integrating Health and Safety into the Design and Operation of Water Systems	To identify and evaluate exemplary practices that integrates health and safety improvements into the design of water systems and effective health and safety management practices into routine operation and maintenance.
3. Establishing the Full Costs of Injuries and Illnesses and Evaluating the Cost/Benefit of Prevention Programs	To identify the full cost of injury/illness (including cost associated with lost work time, overtime, medical costs, lost productivity, etc.). This project was also intended to identify cost/benefit of developing and maintaining various types of prevention programs.
4. National Water Occupational Health & Safety Database: Feasibility and Pilot of Benchmarking Data Consortium	To determine the feasibility of a national occupational health and safety database on injuries and illnesses in the water industry and to develop a scope for a future pilot study to test the functionality and usability of such a unified data structure.
5. Evaluation of the Effectiveness of Alternate Enforcement Methods and Approaches	To identify and evaluate the effectiveness of alternate enforcement methods in improving water utility safety performance.
6. Effective Infrastructure Protection Techniques for Protection of Water System Employees and the Public	To identify and evaluate effective infrastructure protection techniques in preventing injury to water system employees and the public.
7. The Role of Human Factors in Injuries and Improving Safety Performance in the Water Industry	To better understand the effects of human factors on safety performance and to provide guidance to improve the health and safety of water utility workers and prevent potential effects to the public by considering human factors in system design and operations.
8. The Effects of Chronic Low Level Exposures to Airborne Chemicals at Water Utilities	To identify and understand the extent and effects of chronic low level exposure to chlorine, ammonia and/or chloramine vapors by employees in the water industry and to provide guidance on how to minimize/prevent employee exposures to these vapors.

Prioritization of Research Projects

The workshop participants identified seven of the most important criteria that could be used to rank potential research projects. Table 4.4 lists these in the order of importance as determined by each participant. The participants ranked criteria assigning a number between 1 and 5 (with 1 being most important and 5 being least important). The average of the 21 scores submitted was calculated for each criterion.

The averages were then used to develop an importance weighting factor, calculated by the formula:

Weighting Factor = 5 - (Average Importance Rank)

These results were rounded to the nearest whole number. The five most important criteria were then used, along with the weighting factors to rank the potential research projects.

Table 4.4
Importance of Research Project Ranking Criteria

Criteria No.	Criteria	Individual Importance Rank Responses																					Average Importance Rank	Weighting Factor
2	Severity of Event/Impact	2	1	2	2	2	1	2	1	1	1	2	3	1	1	1	1	1	1	1	1	1	1.4	**4**
1	Frequency of Event/Impact	2	1	1	3	2	1	1	2	1	1	2	2	1	1	4	2	1	1	1	2	1	1.6	**4**
3	Ability to Prevent/Improve Outcomes	3	1	3	1	1		2	2	2	1	3	2	3	2	3	1	3	2	1	1	5	2.1	**3**
5	Ability of Industry to Implement Likely Solution (incl. cost/benefit & obstacles)	1	3	2	2	1	2	2	1	3	2	1	1	1	1	4	3	2	3	3	3	5	2.2	**3**
4	Likelihood of Success of R&D in Identifying Solution	2	4	1	1	2	2	2	2	3	1	4	1	3	3	2	4	3	3	2	2	3	2.4	**3**
6	If we don't do it who will?	4	4		5	4	2	4		4	3	1	4	5	4	1	5	2		3	5	1	3.4	**2**
7	Ability to get co-funding	4	3	5	5	4	5	4		5	3	5	3	5	5	3	5	2	5	3	5	5	4.2	**1**

Note: Blank entries, indicating no score provided, were not included in the average.

Based on the weighting factor, the potential research projects were prioritized. The results are presented in Table 4.5. The total score was obtained by adding up all the individual scores, and the overall rank is the percentage of the scores each of the project received relative to the score of the highest ranked project.

Table 4.5
Research Project Ranking

Potential Research Project	Total Score	Overall Rank
1. Effectiveness of Techniques for Prevention of Lower Back and Shoulder Injuries and Other Musculoskeletal Disorders	1464	100
2. Best Practices for Integrating Health and Safety into the Design and Operation of Water Systems	1448	99
3. Establishing the Full Costs of Injuries and Illnesses and Evaluating the Cost/Benefit of Prevention Programs	1436	98
4. National Water Occupational Health & Safety Database: Feasibility and Pilot of Benchmarking Data Consortium	1320	90
5. Evaluation of the Effectiveness of Alternate Enforcement Methods and Approaches	1058	72
6. Effective Infrastructure Protection Techniques for Protection of Water System Employees and the Public	1047	72
7. The Role of Human Factors in Injuries and Improving Safety Performance in the Water Industry	937	64
8. The Effects of Chronic Low Level Exposures to Airborne Chemicals at Water Utilities	931	64

DEVELOPMENT OF PROJECT SCOPES FOR SELECTED PROJECTS

The four highest-ranking potential research projects from Table 4.5 were selected for development of preliminary project scopes during the workshop. The participants were split into two groups and asked to outline an objective, a suggested approach to implementing the project, an estimate of the budget cost, and an estimated time schedule for each of the selected projects.

After the workshop, F-DE&S further refined these project scopes and developed project scopes for the remaining high priority projects. The draft project scopes for each of the high priority research projects are presented in Appendix C.

CHAPTER 5
THE AWWARF H&S R&D AGENDA

NEED FOR AN R&D AGENDA

All water utilities are subject to injuries/illnesses and can benefit from improved safety programs and technologies, even those with exemplary programs and safety performance. The water industry as a whole appears to have higher recordable incidence rates, lost workday incidence rates, and severity rates when compared to other utilities (i.e., electric utilities). In addition, government operated water utilities operate in an environment without health and safety regulations comparable to U.S. OSHA rules in almost half of the states of the U.S. These utilities are more susceptible to complacency and poor performance because regulatory enforcement is often the most potent factor driving health and safety programs.

Current performance suggests that water utility executives and industry associations such as AWWA and AwwaRF should make concerted efforts to improve health and safety performance. Water utilities need not wait to make a commitment to compliance and safety excellence and to make aggressive use of existing safety management tools and techniques.

Organizations like AwwaRF can assist in furthering the goal of reducing injuries and illnesses in the industry by conducting research and development to support performance improvements. An R&D agenda is essential to direct available resources to the most beneficial projects and to provide a basis for receiving co-funding from other entities interested in the same goals. Because R&D funding is limited, this agenda (based upon the most critical needs in the industry and the areas that can most benefit from research) will yield the most benefits for the industry.

RECOMMENDED PROJECTS

Eight critical research projects were identified by the participants in this workshop and prioritized based upon both need and likelihood of success in achieving safety performance improvement, as described in Section 4. These projects, in order of priority, are:

1. Effectiveness of Techniques for Prevention of Lower Back and Shoulder Injuries and Other Musculoskeletal Disorders
2. Best Practices for Integrating Health and Safety into the Design and Operation of Water Systems
3. Establishing the Full Costs of Injuries and Illnesses and Evaluating the Cost/Benefit of Prevention Programs
4. National Water Occupational Health & Safety Database: Feasibility and Pilot of Benchmarking Data Consortium
5. Evaluation of the Effectiveness of Alternate Enforcement Methods and Approaches
6. Effective Infrastructure Protection Techniques for Protection of Water System Employees and the Public
7. The Role of Human Factors in Injuries and Improving Safety Performance in the Water Industry

8. The Effects of Chronic Low Level Exposures to Airborne Chemicals at Water Utilities

AwwaRF research and development is devoted to achieving a wide variety of objectives and funding for any one of these objectives is limited. F-DE&S recommends that the goal of reducing injuries/illnesses be among the top AwwaRF priorities. F-DE&S recommends that a minimum of 1-2 health and safety projects be funded in each year. With that level of AwwaRF commitment, the above list represents an R&D agenda for the 5-10 year horizon.

POTENTIAL PARTNERS

Participants in the workshop identified potential research partners who could potentially share in the funding and/or execution of these research projects. A brief description of the organizations and the benefits of partnering with them follow.

<u>National Institute for Occupational Safety and Health (NIOSH)</u>

NIOSH is responsible for funding, directing and conducting research on the full scope of occupational disease and injury ranging from lung disease in miners to carpal tunnel syndrome in computer users. In addition to conducting research, NIOSH:

- investigates potentially hazardous working conditions when requested by employers or employees;
- makes recommendations and disseminates information on preventing workplace disease, injury, and disability; and
- provides training to occupational safety and health professionals.

NIOSH has developed a National Occupational Research Agenda and is actively funding projects and seeking funding partners for projects that meet the objectives of its agenda. The above recommended AwwaRF projects are believed to be consistent with the NORA objectives and may be candidates for NIOSH co-funding.

<u>Department of Labor's Bureau of Labor Statistics (BLS) and Occupational Health and Safety Administration (OSHA)</u>

BLS is the principal fact-finding agency for the Federal Government in the field of labor economics and statistics and has recently studied performance data for government employees. OSHA is responsible for regulatory enforcement and a recent Department of Labor’s Office of Inspector General report described in Section 4 recommended that they do more to encourage and promote state programs in non-OSHA Plan States. While they do not have major R&D budgets, BLS and OSHA would be effective partners for projects 4 and 5 addressing improved water industry data collection and alternative enforcement methods and approaches. Their involvement in these projects could provide technical assistance and attract other funding.

US Environmental Protection Agency

The U.S. Environmental Protection Agency (EPA) Office of Water funds research and development devoted to ensuring that people are provided with safe drinking water. In the past, it has funded related research regarding safety issues in the water industry, including the development of the AwwaRF *Compliance Guidance and Model Risk Management Program for Water Treatment Plants* and the development of vulnerability assessment tools to assist the water industry in preventing or minimizing the effects of terrorist attack. EPA may be a potential partner in co-funding for all of the identified projects, especially projects 2, 6 and 8.

State Workers' Compensation Organizations

State workers compensation organizations have goals that are consistent with AwwaRF's safety performance improvement goals. Some States have had research groups as part of their workers compensation programs, like Florida's Division of Workers Compensation, Bureau of Research and Education, a project participant. Some organizations may be willing to partner and/or co-fund projects, especially projects 3 and 4, which are focused on data collection and promoting understanding of the full costs of injuries and illnesses.

Professional and Industry Associations

Several professional organizations were identified as having a strong common interest in safety performance improvement.

The National Safety Council, a project participant, does not generally fund such research, but is committed to promoting and participating in such critical efforts. Their involvement could be especially helpful for project 4.

The International Association of Industrial Accident Boards and Commissions (IAIABC) is an association of workers compensation board members and professionals that is an ideal umbrella organization to involve state workers compensation organizations.

The Water Environment Research Foundation (WERF), associated with the Water Environment Federation (WEF), funds research critical to wastewater treatment operations. WERF is a natural partner for this effort because wastewater collection and treatment operations most faced with identical safety issues as the water industry.

Private Research Institutions and National Insurance Companies

Private research institutions, including those operated and funded by insurance companies and universities, may be potential partners in co-funding and performing the research. National insurance companies may also be approached to fund research in these fields.

Water Utilities

As with all AwwaRF projects, interested water utilities may participate in projects and may also co-fund projects that directly address issues that affect them. Potential partners include those organizations that have had to re-examine their safety performance and culture due to fatalities or regulatory enforcement actions.

WHAT WATER UTILITIES CAN DO RIGHT NOW

The goals of this project were to develop a research and development agenda and high priority R&D project scopes for the water industry. Nevertheless, the work performed and this report provides important insights to water industry executives and health and safety staff regarding what they can do to improve safety programs and performance and where to best focus their resources.

The following is a brief summary of lessons learned from the project, the workshop and the participating health and safety specialists from the water industry and other stakeholder organizations.

Top 10 Things Water Industry Executives Can Do

1. Develop a H&S policy or mission statement. Actively and regularly communicate that working safely and protecting employees is a core value of the organization.
2. If you are a government operation and not in an OSHA-Plan State (see Section 4.2.2), make a policy commitment for your organization to comply with OSHA regulations.
3. Develop a Health and Safety Oversight Committee of key department managers to review safety performance, set goals, and direct safety improvement programs.
4. Track the full costs of injuries and illnesses to the organization.
5. Integrate safety goals into performance review criteria from executives to supervisors.
6. Create a process to identify and give high priority to safety-critical procurement/project requests and to have safety review of major projects.
7. Provide sufficient and well-trained safety specialists and provide the financial resources necessary to support an effective program (including time for training).
8. "Walk-the-walk" by periodically attending facility safety meetings and conducting field safety walkthroughs with health and safety and supervisory staff.
9. Hold supervisors accountable for ensuring their employees comply with safety rules.
10. Provide a work environment that encourages all employees to communicate safety concerns without retribution to their supervisors and, if necessary, to impartial parties.

Top 10 Things Health and Safety Staff Can Do

1. Involve executive management in Health and Safety and regularly engage them in discussions regarding performance, new initiatives and the need for involvement.
2. Develop and implement formal, written safety management programs.
3. Provide training to meet both regulatory requirements and to enhance performance.
4. Develop a network of safety representatives or leaders that reaches all parts of the organization to allow effective communication in both directions. Develop safety committees in each substantial facility or operating group to involve employees in identifying and solving problems close to where they occur.
5. Collect injury/illness data in way that provides for breakdown by department, location and type of work. Analyze data to track performance vs. goals and history, benchmark

with peers, and breakdown by department/type of work to focus efforts on problem areas.
6. Conduct meaningful investigations, with employee involvement, of safety incidents that includes assessment of root causes and contributing causes.
7. Investigate "near-misses", incidents that could have easily resulted in injury if the employee were not so lucky, not just injuries. Take action before an injury occurs.
8. Implement safety recognition and reward programs that encourage good practices and performance without suppressing reporting and investigation.
9. Conduct periodic audits of operations for compliance with OSHA regulations and your organization's policies and programs.
10. Provide clear guidelines for injury classification and management from immediate care through rehabilitation and return to work.

Top 10 Things Supervisors Can Do

1. Lead by example, following all safe work practices and rules.
2. Regularly communicate that safety cannot be sacrificed and getting the job done means getting it done safely.
3. Ensure that adequate protective gear and safety supplies are available to employees.
4. Discuss how to get each hazardous job done safely while setting up the work.
5. Periodically review the work site for safety and compliance with safe work practices and rules.
6. Convey interest in understanding and addressing problems identified by employees.
7. Don't dismiss any safety concerns raised by employees. Carefully explain yourself if there is a concern that you believe is unfounded or obtain the opinion of H&S staff.
8. Get actively involved in solving safety problems as they arise on the job.
9. Be receptive to changes in H&S programs and safe work practices.
10. Actively participate in training opportunities and other safety opportunities that arise.

Top 10 Things Employees Can Do

1. Always follow safe work rules and procedures.
2. Mentally go over safety hazards and requirements at the start of every job.
3. Wear protective gear appropriate to the job task.
4. Alert your co-workers if you notice them putting themselves and others at risk of injury by not following procedures.
5. Don't sacrifice safety just to get the job done sooner.
6. Communicate safety concerns to supervisors or, if necessary, through other available means provided by your organization.
7. Get actively involved in solving safety problems as they arise on the job.
8. Don't raise knowingly false concerns over safety, this can make it more difficult for your organization to address important safety concerns and measures.
9. Be receptive to changes in H&S programs and safe work practices.
10. Actively participate in training opportunities and other safety opportunities that arise.

APPENDIX A

DEFINITIONS FOR INCIDENCE DATA REPORTED

DEFINITIONS FOR INCIDENCE DATA REPORTED

The standard definitions used for reporting injuries and illnesses are summarized as follows:

Recordable occupational injuries or illnesses are any occupational injuries or illnesses that result in:
(1) Fatalities, regardless of the time between the injury and death, or the length of the illness; or
(2) Lost workday cases, other than fatalities, that result in lost workdays; or
(3) Nonfatal cases without lost workdays which result in transfer to another job or termination of employment, or require medical treatment (other than first aid) or involve: loss of consciousness or restriction of work or motion. This category also includes any diagnosed occupational illnesses that are reported to the employer but are not classified as fatalities or lost workday cases.

Lost workdays: The number of days (consecutive or not) after, but not including, the day of injury or illness during which the employee would have worked but could not do so; that is, could not perform all his normal assignment during all or any part of the workday or shift, because of the occupational injury or illness.

OSHA No.200[1], Log and Summary of Occupational Injuries and Illnesses: Each recordable occupational injury and illness must be logged on this form within six working days from the time the employer learns of it. If the log is prepared at a central location by automatic data processing equipment, a copy current to within 45 calendar days must be present at all times in the establishment. A substitute for the OSHA No.200 is acceptable if it is as detailed, easily readable and understandable as the OSHA No.200.

OSHA No.101, Supplementary Record of Occupational Injuries and Illnesses: The form OSHA No.101 contains much more detail about each injury or illness. It also must be completed within six working days from the time the employer learns of the work-related injury or illness. A substitute for the OSHA No.101 (such as insurance or workers' compensation forms) may be used if it contains all required information.

Incidence Rates: The following rates are generally used based on the exposure of 100 full-time workers using 200,000 employee-hours as the equivalent (100 employees working 40 hours per week for 50 weeks per year):

$$\text{Recordable Incidence Rate} = \frac{\text{No. Of Total Cases of Recordable Events} \times 200{,}000}{\text{Total hours worked by all employees during period covered}}$$

$$\text{Lost Workday Incidence Rate} = \frac{\text{No. Of Recordable Injuries causing lost days} \times 200{,}000}{\text{Total hours worked by all employees during period covered}}$$

$$\text{Severity Rate}^2 = \frac{\text{No. Of Lost Workdays (after day of injury)} \times 200{,}000}{\text{Total hours worked by all employees during period covered}}$$

---

[1] Note: The OSHA 200 log was in use during the period this project was conducted. This has been replaced with the OSHA 300 log.

[2] This is the incidence rate of lost workdays as defined by BLS and OSHA (also called Lost Workdays Incidence Rate).

APPENDIX B

F-DE&S HEALTH AND SAFETY SURVEY

AwwaRF Research Foundation Project 2684

Health and Safety Survey of Water Utilities to Support R&D Agenda Workshop

THIS SURVEY WILL HELP US to help you and others in the water industry improve safety and loss performance. Water supply systems around the world are subject to the effects of work-related accidents and environmental health-related illnesses. Injuries and illnesses can result in lost productivity, increased worker's compensation cost, and poor employee morale. Regulatory non-compliance can result in expensive penalties and reduced consumer confidence in the utility.

OUR ULTIMATE GOAL is identify emerging Health and Safety Issues in the water industry and to develop a Research and Development agenda to allow the Awwa Research Foundation to address the most pressing industry needs. Ultimately, this will allow you to avoid major losses, and improve your ability to minimize incidents, injuries and illnesses that affect your employees and organization.

IN ORDER TO ENSURE that we meet our goals, *we need your help!* We are asking you to complete a survey on lessons learned from major disasters you have experienced.

YOU MAY BE SELECTED TO REPRESENT THE INDUSTRY AT A WORKSHOP! By sharing your experiences and identifying critical needs and exemplary practices of your organization, you may be receive an invitation to attend a two day workshop in Fort Lauderdale to review critical needs and develop an R&D agenda for the industry. Selected responders who can share important experiences and perspective with us will have expenses paid to the workshop.

YOU MAY NEED THE HELP OF OTHER health and safety and operations staff. If you cannot answer questions, just explain why and move on to the next one. Please submit a response *with whatever questions you can answer.*

SEND A HARD COPY OR ELECTRONIC RESPONSE BY SEPTEMBER 15TH TO:

Mr. Peter S. Puglionesi
Framatome ANP Duke Engineering & Services
457 Haddonfield Rd., Suite 200
Cherry Hill, NJ 08002

Phone: 856- 910-0097 or (610) 449-3430
Fax: 856- 910-7029
email: pspuglio@dukeengineering.com

THANK YOU FOR PARTICIPATION in this project, which is critical to the continued health and safety of our industry and to the welfare of our employees!

HEALTH AND SAFETY SURVEY OF WATER UTILITIES

YOUR NAME:	PHONE NUMBER:
YOUR TITLE:	EMAIL ADDRESS:
ORGANIZATION NAME:	
ORGANIZATION ADDRESS:	
CITY, STATE, ZIP CODE:	

GENERAL INFORMATION

The purpose of this section is to identify the different types of systems in the water industry and the general approaches used for Health and Safety.

1. Type of utility (check ALL that apply):
   ☐ Water ☐ Waste water ☐ Electricity
   ☐ Gas ☐ Other (please specify):

2. Total number of employees in 2000:

3. Total number of productive (i.e., at work) hours in 2000:

4. Total population served in 2000: Retail ______ Wholesale (if any)

5. Check which one of the following applies to your organization (check only one):
   ☐ Government owned and operated ☐ Government owned but privately operated
   ☐ Privately owned and operated

6. Your understanding is that your organization is (check only one):
   ☐ Subject to Federal OSHA regulations
   ☐ Exempt from Federal OSHA regulations but subject to State regulations that are substantially equivalent to Federal OSHA regulations
   ☐ Exempt from Federal OSHA regulations and there are NO State regulations that are substantially equivalent to Federal OSHA regulations

7. If you checked the last box in Question 6, what best describes your organization's approach regarding OSHA regulations:
   ☐ We are not subject to OSHA regulations and do not follow them
   ☐ We recognize that OSHA regulations may be considered a standard of care and we try to follow them because we think they are important to safe operation, although not every regulation
   ☐ We try to follow all OSHA regulations
   ☐ Executive management has made a formal policy commitment to follow OSHA regulations
   ☐ We are committed to follow OSHA regulations and we conduct regular audits of our program and facilities against OSHA regulations
   ☐ We are committed to meet and go well beyond OSHA regulations
   ☐ Other (describe):

8. Have any of your facilities participated in the OSHA Voluntary Protection Program?
   ☐ Yes ☐ No ☐ Don't know

9. How does your organization pay for workman's compensation benefits?
   ☐ Self-Insured/Self-Administered ☐ Self-Insured/Contract-Administered
   ☐ Commercial Insurance Policy ☐ Publicly Operated State Insurance Fund

HEALTH AND SAFETY PROGRAM ROLES AND ACCOUNT-ABILITIES

The purpose of this section is to characterize Health and Safety management approach and commitment.

10. Check ALL of the methods your organization uses to define Health and Safety performance expectations and determine performance:
    - ☐ Executive management sets Health and Safety Policy
    - ☐ The organization has measurable Health and Safety (H&S) performance goals
    - ☐ Executives, managers, supervisors and other employees are held accountable for meeting non-measurable H&S performance goals in performance evaluations
    - ☐ Executives, managers, supervisors and other employees are held accountable for meeting measurable H&S performance goals in performance evaluations
    - ☐ Other (describe):

11. Check ALL of the methods your executive management uses to communicate their interest in Health and Safety:
    - ☐ Executive management delegates these responsibilities to staff responsible for Health and Safety
    - ☐ Executive management helps to troubleshoot problems after severe injury cases
    - ☐ Executive management includes Health and Safety (H&S) programs and performance as a regular agenda item in staff and management meetings
    - ☐ Executives participate in H&S meetings on a regular basis
    - ☐ Other (describe):

12. Who has overall responsibility for Health and Safety program oversight and performance monitoring?
    - ☐ Safety Manager
    - ☐ Human Resources Manager
    - ☐ Operations Manager
    - ☐ Individual Supervisors
    - ☐ Engineering
    - ☐ Safety Committee
    - ☐ Other (please specify):

    Is this responsibility defined in a writing?
    - ☐ Yes
    - ☐ No

13. What best describes the duties of the person with overall responsibility for Health and Safety program oversight and performance monitoring:
    - ☐ Dedicated Position with No Other Responsibilities
    - ☐ Primary Focus on Health and Safety with other minor duties
    - ☐ Primary Focus on other duties with minor focus on Health and Safety
    - ☐ Other (please specify):

14. What best describes your organization's use of trained or certified Health and Safety specialists:
    - ☐ Degreed/certified H&S specialists on staff
    - ☐ H&S specialists on staff; degree or certification not required
    - ☐ H&S specialists normally accessed through insurer
    - ☐ H&S specialists normally accessed through standing contract
    - ☐ Do not normally use H&S specialists
    - ☐ Other (please specify):

HEALTH AND SAFETY PROGRAM ATTRIBUTES

15. Does your organization have written Health and Safety Programs?
☐ Yes ☐ No

If yes, the Health and Safety Program was (check all that apply):
☐ Purchased as a Generic Safety Program and Not Customized
☐ Purchased as a Generic Safety Program and Customized
☐ Developed by Corporate Office for Use at Multiple facilities
☐ Custom Safety Program Developed by facility personnel
☐ Custom Safety Program Developed by outside vendor

Management Systems

16. What is your organization's approach to formal Health & Safety (or EH&S) management systems?
☐ We are not considering development of a formal H&S management system at this time
☐ We are planning to develop a formal H&S management system
☐ We have a formal and well documented H&S management system that we think meets the intent of programs like ISO 14000
☐ We plan to obtain ISO 14000 certification for our H&S management system
☐ We have obtained ISO 14000 certification for our H&S management system

Budget

17. Does your organization track Loss and Health and Safety Costs?
☐ Yes ☐ No
If Yes, briefly describe how loss and health & safety costs are broken down and tracked across the organization. (e.g., by operating division, job classification, department, etc.)?

Please Provide a Brief Breakdown of a recent year's loss incident related Costs *(Tracked as noted below* ☐ *Rough Estimate* ☐*)*

Note: Skip breakdown if no information is available

Medical Expenses	$
Wage and Productivity Loss	$
Motor Vehicle Damage	$
Property Damage	$
Fire Losses	$
Administrative Costs (investigation, hiring, OT, etc)	$
Other: ______	$

Please Provide a Brief Breakdown of a recent year's annual Health and Safety Program Costs

Labor Cost	$
Medical Exams/Testing	$
Training Materials/Equipment	$
Safety Supplies	$
Professional Services	$
Incentive Programs/Awards	$
Other: ______	$
Other: ______	$

Safety Communication / Employee Participation)

18. Does your organization have a written program for Safety Communications and Employee Participation?
    ☐ Yes ☐ No
    If yes, briefly describe the process(es) for communicating health & safety information to employees?

    Briefly describe the process for employees to communicate health & safety information to the organization?

19. Does your organization have Safety Meetings?
    ☐ Yes ☐ No
    If yes, how often?
    If yes, who is the highest ranking supervisor/manager that regularly attends the meetings?
20. Does your organization have a Safety Committee?
    ☐ Yes ☐ No
    If yes, how often does the committee meet?
    If yes, who is the most highest ranking member?
    How are meeting minutes or committee discussions communicated to other employees?

21. Does Your Organization have (*breifly describe each that applies*):
    21a. Specific incentives (if any) for employers to introduce measures that will decrease the frequency and severity of accidents:

    21b. Specific incentives (if any) for employees to follow safe work practices:

    21c. Specific incentives (if any) given to disabled workers so that they return to productive employment as soon as possible:

Hazard Assessment & Control

22. Which of the following workplace hazard assessment techniques does your organization use? (check all that apply)
    ☐ Conduct PPE and other OSHA defined hazard assessments
    ☐ Conduct Inspections of Workplace; How Often?
    ☐ Conduct Job Safety Analysis (e.g., Nat'l. Safety Council method)
    ☐ Review Safety & Health Information (workers comp., injury data, etc.) How Often?
    ☐ Evaluate new equip., materials, processes before they are introduced
    ☐ Other (please specify):
    ☐ None

23. Does your organization document identified problems and track completion of corrective actions?
    ☐ Yes ☐ No

    Briefly describe your method for tracking corrective actions.

Training 24. Do employees receive Health and Safety training?

☐ Yes ☐ No (skip to no. 30)

If yes, How many hours/yr for a typical employee?

What is your organization's principle training method? (give approximate % if you check more than one)

☐ Classroom (____%)
☐ On-the-Job (____%)
☐ Computer-based (____%)
☐ Self-Study (____%)
☐ Other (please specify): ______ (____%)

25. Which of the following apply to the administration of your organization's health & safety training? (check all that apply)

☐ Dedicated Health & Safety Training Coordinator
☐ Health & Safety Training is Included with other employee training
☐ Required training is tracked and scheduled manually
☐ Required training is tracked and scheduled with computer software
☐ Employee attendance in health & safety training is documented
☐ Testing or some method of determining comprehension or skill is used. Please Specify Method

26. How are employee participation to Health and Safety training promoted (check all that apply)?

☐ Training is mandatory, unless there is a scheduling problem
☐ Training is mandatory, and completion of initial and refresher training is always completed, by re-scheduling if necessary
☐ Employee incentives tied to health & safety training
☐ Other (please specify):

27. Which programs are covered in the training (check all that apply)?

☐ Bloodborne Pathogens ☐ Boiler Safety ☐ Compressed Gas
☐ Confined Spaces ☐ Crane & Hoist Safety ☐ Electrical Safety
☐ Egress Program ☐ Emergency Response ☐ Ergonomics
☐ Excavation/Trenching ☐ Fall Prevention/Protection ☐ Fire Safety
☐ First Aid ☐ General Safety Policies ☐ Hazard Communication
☐ Hearing Conservation ☐ Heat Stress Prevention ☐ Hot Work
☐ Housekeeping ☐ Lockout – Tagout ☐ Machine Guarding
☐ Office Safety ☐ Personal Protective Equip ☐ Radiological Safety
☐ Respiratory Protection ☐ Scaffolds/Ladders/Work Surfaces
☐ Specialty Chemicals (i.e. Asbestos, Pesticide)
☐ Portable & Hand Tool Safety
☐ Visitor Safety/Orientation
☐ Other (please specify):
☐ Other (please specify):
☐ Other (please specify):

Incident Investigation

28. Does your organization have a written incident investigation procedure?

- ☐ Yes (please attach, if possible) ☐ No

If Yes, which of the following are attributes of the procedure and practice? (check all that apply)

- ☐ Requires Investigation of Accidents/Injuries Only
- ☐ Requires Investigation of "Near Misses", and they are normally done
- ☐ Requires Investigation of "Near Misses", not normally done
- ☐ Supervisors encouraged to Identify & Investigate "Near Misses"
- ☐ Requires a trained leader of the investigation team
- ☐ Encourages a root cause analysis
- ☐ Requires a root cause analysis
- ☐ Focuses on corrective actions
- ☐ Requires communication of findings/recommendations
- ☐ Requires safety committee review
- ☐ Other (please specify):

Did you attach a copy of your incident investecation procedure? ☐

Health & Safety Recordkeeping

29. Your organization collects injury data through (check all that apply):

- ☐ Periodic reports (please specify):
  - ☐ Monthly summary of injuries and illnesses
  - ☐ Annual report
  - ☐ OSHA No. 200 or OSHA No. 101 [a]
  - ☐ ANSI Z16.5 [b]
  - ☐ Other (specify):
- ☐ First Aid Report [c]
- ☐ Injury and illness record of employee [d]
- ☐ Incident Investigation Report [e]
- ☐ Near Miss Report

[a] *OSHA recordkeeping system mandatory for all establishments subject to the OSHAct.*

[b] *American National Standards Institute.*

[c] *Report filled out by the first aid attendant or nurse at the time an injured or ill person comes for treatment.*

[d] *Records injury experience of individual employees.*

[e] *Provides a record of contributing circumstances to provide a basis for specific remedial actions*

> Definitions as per OSHA and other guidance for completing the Injuries/Illnesses table on the following page.
>
> Recordable occupational injuries or illnesses are any occupational injuries or illnesses that result in:
> (1) Fatalities, regardless of the time between the injury and death, or the length of the illness; or
> (2) Lost workday cases, other than fatalities, that result in lost workdays; or
> (3) Nonfatal cases without lost workdays which result in transfer to another job or termination of employment, or require medical treatment (other than first aid) or involve: loss of consciousness or restriction of work or motion. This category also includes any diagnosed occupational illnesses that are reported to the employer but are not classified as fatalities or lost workday cases.
>
> Lost workdays: The number of days (consecutive or not) after, but not including, the day of injury or illness during which the employee would have worked but could not do so; that is, could not perform all his normal assignment during all or any part of the workday or shift, because of the occupational injury or illness.
>
> Number of Full-Time Employees: If the departments in your organization do not break down the work intor the categories given on the table, provide a rough estimate of the number of employees performing the function, but make sure that the total number of employees is correct for your organization. If you have a significant number of part-time staff, use Full-Time Equivalents (total people-hours/week divided by 40) rounded to the nearest whole number.
>
> NOTE: Please skip entries where no data is available for your organization.

Notes: *ONLY INCLUDE WATER SYSTEM EMPLOYEE /INCIDENTS*

INJURIES / ILLNESSES AND LOST WORKDAYS	DEPARTMENT / OPERATION:		(A) # CASES WITH NO LOST WORK-DAYS	(B) # CASES WITH LOST WORK-DAYS	(C) # OF LOST WORK-DAYS	(D) # OF FULL-TIME EMPLOY-EES
Include accident data ONLY from water-associated activities.	30. SOURCE OF SUPPLY	*Employees working at surface reservoirs, collection systems, etc.*				
		Employees working at groundwater reservoirs, collection systems, etc.				
	31. WATER TREATMENT	*Employees (i.e., operating personnel) working in a surface water treatment facility*				
		Employees (i.e., operating personnel) working in a groundwater treatment facility				
	32. ADMINISTRATION / CUSTOMER SERVICE *Employees working in an office environment*					
	33. DISTRIBU-TION	*Construction of pipelines, hydrants, etc. in distribution systems*				
		Maintenance of pipelines, hydrants, etc. in distribution systems (not meters)				
	34. LABORATORY *Personnel working in water quality laboratories*					
	35. MAINTEN-ANCE	*Employees involved in facility and equipment maintenance (not pipeline or meter maintenance)*				
	36. METER *Employees actually reading meters (not billing department); meter installation/repair also included*					
	37. TOTAL COLUMN A: INJURIES/ILLNESSES WITHOUT LOST WORKDAYS: *Total number of cases that required medical treatment (not merely first aid) but which did NOT result in restricted days or days away from work.*		ENTER (A) TOTAL HERE:			
	38. TOTAL COLUMN B: LOST WORKDAY CASES: *This is the total number of cases that occurred when an injured or ill employee experienced days away from work. Must affect employee beyond day of injury or onset of illness.*			ENTER (B) TOTAL HERE:		
	39. TOTAL COLUMN C: LOST WORKDAYS: *This is the total number of lost workdays due to occupational injuries or illnesses in 2000*				ENTER (C) TOTAL HERE:	
	40. TOTAL COLUMN D: FULL-TIME EMPLOYEES: *This is the total number of full-time employees (or equivalents, if significant part time work) in 2000.*					ENTER (D) TOTAL HERE:
	41. TOTAL NEAR MISSES INVESTIGATED *This is the total number of Near Misses investigated that were NOT reportable cases (and not listed above) for the year 2000:*					

INJURIES / ILLNESSES AND LOST WORKDAYS	LOST WORKDAY INJURY/ ILLNESS TYPE AS A RESULT OF:	# of lost work days in 2000
Include accident data ONLY from water-associated activities.	42. Trenching cave-in	
(continued)	43. Confined space entry	
	44. Motor vehicle accident	
	45. Heavy equipment accident	
	46. Chemical exposure	
	47. Slips and falls	
	48. Insect/animal bites	
	49. Back injuries	
	50. Other injuries	

MOTOR VEHICLE ACCIDENTS

Any occurrence involving a utility motor vehicle resulting in death, injury, or property damage, unless such vehicle is properly parked, should be included.

51. Number of water system vehicles:
52. Total miles driven during 2000:
53. Total number of accidents:
54. Type of Accident (if breakdown is available):
    - # of head-ons
    - # at an angle
    - # of backing
    - # of vehicle behind
    - # of vehicle ahead
    - # of involving pedestrian
    - # of sideswipes
    - # of fixed object
    - # of miscellaneous

CLASSIFICATION OF INJURIES / ILLNESSES

55. Does your organization maintain injury/illness data based on:
    ☐ Department ☐ Job Classification ☐ Job Task
    ☐ Other (list)

56. Does your organization maintain injury/illness data based on:
    ☐ Nature of injury/illness (fracture, sprain, etc.)
    ☐ Initiating Event or Exposure (fall, overexertion, etc.)
    ☐ Other (list)

57. What would you consider to be the 2 Departments in your organization that are involved in the Most Hazardous work? (Please List with a Brief Explanation of the Hazards and rough # of employees involved)
    1. ______-
    2. -

58. What would you consider to be the 5 Job Classifications in your organization that are perform the Most Hazardous work? (Please List with a Brief Explanation of the Hazards and rough # of employees involved)
    1. ______-
    2. ______-
    3. ______-
    4. ______-
    5. ______-

59. What would you consider to be the 5 Specific Job Tasks or Activities that you consider to be the Most Hazardous for your employees? (Please List with a Brief Explanation of the Hazards and rough estimate of manhours/year spent on the activity)

1. ______-
2. ______-
3. ______-
4. ______-
5. ______-

Please answer the following questions to help us identify the most critical needs and develop an R&D agenda to help improve safety performance in the water industry.

HEALTH & SAFETY RESEARCH AND DEVELOP-MENT NEEDS

60. What would you consider to be the top 5 needs of your organization with regard to the overall health & safety of your employees?

1. ______-
2. ______-
3. ______-
4. ______-
5. ______-

61. What research and development projects could help address the needs identified above? Please Be Specific

62. Do you have a health and safety program or initiative that you feel has been very effective in reducing injury/illness or other losses?

☐ Yes ☐ No

If Yes, please describe your program or attach a program description.

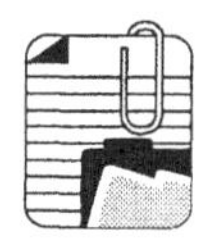

Did you attach a copy of your exemplary Health and Safety Program?

☐

Do you feel that you have either exemplary Health & Safety program experiences to present or can otherwise greatly contribute to the development of a Health & Safety R&D agenda for the water industry in a Workshop setting (note, expenses will be paid for selected responders to attend a 2 day workshop to develop a research and development agenda)?

☐ Yes ☐ No

If Yes, please explain why

APPENDIX C

POTENTIAL RESEARCH PROJECT SUMMARIES

Research Project 1: Effectiveness of Techniques for Prevention of Lower Back and Shoulder Injuries and Other MSDs

Objective

The objective of this research will be to identify the most effective techniques for preventing lower back, shoulder and other MSDs in the water industry. The research will identify the most cost-effective strategies for reducing the high cost of these most common and costly injuries to water systems and society and identify potential improvements that are most effective in water industry applications.

Background

Lower back, shoulder and MSD injuries consume a large amount of organizational resources. Lower back disorders are one of the largest causes of lost time and workers' compensation claims in the United States and in the water industry. About $11 billion and $2.1 billion in workers' compensation costs incurred annually for musculoskeletal disorders of the lower back and upper extremities, respectively. Estimates of the total cost of lower back pain to society in 1990 were between $50 billion and $100 billion per year*. Lower back, shoulder and other MSD injuries can be attributed to:

- Material handling/lifting
- Improper body mechanics
- Awkward positions
- Repetitive motion
- Environmental conditions (mud, temperature, unsecured footing, uneven terrain)
- Tool design
- Workstation design (e.g. administrative areas, meter shops, field use of laptops, etc.)
- Position of objects to be lifted (above/below waist)
- Physical condition of employee
- Effects of vibration
- Duration of particular tasks
- Force required vs. force applied

Although items such as backbelts, hydraulic digging, field hoists, etc. are used to address specific problems listed above, their use is not universal and their effectiveness has not been evaluated in the water industry to develop ways of improving them.

The AWWA Health and Safety Committee has been charged with the responsibility of taking a lead role in assembling and disseminating timely state-of-the-art information to increase the

* National Occupational Research Agenda, "Low Back Disorders." Online. www.cdc.gov/niosh/nrlowbck.html, Dec. 4, 2001.

health and safety of water utility workers and affected public. On its own, however, the committee cannot collect, analyze, and report statistically significant data on lower back, shoulder and mulculoskeletal disorder/injury/illness; identify techniques for preventing these problems; and evaluate the effectiveness of these prevention techniques.

Research Approach

The research project will collect and analyze data from water systems, consult with industry/subject matter experts and:

- identify back, shoulder, repetitive motion, and other injuries and illness of the lower back and upper extremities,
- identify the major root cause of these injuries
- identify the use of best prevention practices through evaluating successful organizations,
- evaluate the effectiveness of these best practices in reducing injuries in the water industry
- evaluate current applicable research/information from other industries,
- validate and communicate the results and recommendations for measures that can be directly applied in the water industry,and
- identify potential improvements in mitigation technologies and practices that may require further research to develop or to demonstrate effectiveness in the water industry.

A final project summary will include results conclusion and recommendations from the collection and analysis of the data obtained, and consultation with industry/subject matter experts on the issues listed above.

Project Advisory Committee

Project Advisory Committees (PACs) are organized by the AWWA Research Foundation (AwwaRF) for each funded project to provide guidance, review all reports and significant materials, and generally monitor project performance in behalf of AwwaRF and the water utility industry.

Quality Assurance

Each proposal must include a description of the procedures that will be used to ensure the quality of the data for the project. If the project involves laboratory analyses, this description should indicate whether the laboratory performing the analyses is accredited or state certified for the particular analysis. If the laboratory is not certified, and/or nonstandard methods are used, detailed quality assurance/quality control procedures must be submitted with the proposal.

Budget and Time Schedule

The maximum AwwaRF funding expected for this project is $150K. A minimum 25 percent of the total project cost must be contributed by the contractor. Therefore, the total project cost is $200K ($150K in AwwaRF funds and $50K in contractor contribution). This contribution can either be direct funding or in-kind matching of such items as personal costs, analytical and support services, facilities, consulting services, etc. The submitting organization may elect to contribute more than 25 percent to the project but AwwarF's maximum contribution remains fixed at $150K.

The project period should be realistic, anticipate possible starting delays, and provide ample time for writing, formatting, and revising of the final report and for PAC review of project results. Progress reports will generally be required on a periodic basis (every four months). The final report must be completed in accordance with the AwwaRF's *Format-Style Guide for Preparing Research Report*, unless other arrangements have been mutually agreed upon. The final report should include a separate chapter on recommendations to utilities. Independent of this contract, AwwaRF will fund the PAC as described above.

Research Project 2:
Best Practices for Integrating Health and Safety into the Design and Operation of Water Systems

[*Note: there is some uncertainty over whether this project should focus on the engineering design stage or extend into the life cycle of operation and maintenance. O&M items are identified in italics to allow a change of approach, as appropriate.*]

Objective

The objective of this research will be to identify and evaluate exemplary practices for integrating health and safety improvements into the design of water systems *and integrating effective health and safety management practices into routine operation and maintenance.*

Background

Injury and illness rates for water utilities are higher than for other utilities (i.e., power and natural gas utilities). New or modified water treatment plants, conveyance systems, storage and distribution systems are designed every day and, all to often, water system operators feel that health and safety issues are not adequately considered in the design. This leads to either expensive corrective measures after commissioning or just living with a system that does not facilitate safe work and has greater inherent hazards.

Designing inherently safer systems is an approach that has gained wide acceptance in the petrochemical industry over the last 10 years. This approach is founded on the understanding that the best opportunity for integrating inherent health and safety protections into water treatment plants, conveyance systems, storage and distribution systems is at the design stage. The earlier in the design, the greater the improvements that can be made and the lower the cost.

There is also a growing recognition of the need to re-evaluate health and safety throughout the life cycle of a facility. Appropriate exemplary practices applied during the operations and maintenance of these systems should also be explored.

With ever changing technology, increasing regulatory standards for health and safety, and increasing emphasis on protecting employees and the public, there is a need to identify and communicate exemplary practices to improve health and safety performance during the design *and operation* of water systems. A good understanding of exemplary practices will enable designers and owners of other water systems to benchmark and improve their own approaches to designing *and operating* their systems. Integrating safety into design, *identifying the best safety management practices, and identifying and evaluating the effectiveness of different type of training programs/safety culture (e.g. vehicle and equipment operation training, vendor support training) are some examples of areas in design and operation* of water systems that warrants focus.

Research Approach

The research will identify and evaluate existing practices for integrating health and safety into the design *and operation* of water systems. Methods may include conducting a survey to collect further information on successful approaches, and conducting a workshop for identifying best practices to gain input from experts of the various fields in the water industry.

The final project report will include a description of exemplary practices in the water industry for the design *and operation* of water systems and advice on how to develop and implement practices at water utilities. The report may also include examples of lessons learned in these organizations through past experiences that led to the development of the exemplary practices and recommendations on design and operation approaches to enhance inherent health and safety.

Project Advisory Committee

Project Advisory Committees (PACs) are organized by the AWWA Research Foundation (AwwaRF) for each funded project to provide guidance, review all reports and significant materials, and generally monitor project performance in behalf of AwwaRF and the water utility industry.

Quality Assurance

Each proposal must include a description of the procedures that will be used to ensure the quality of the data for the project. If the project involves laboratory analyses, this description should indicate whether the laboratory performing the analyses is accredited or state certified for the particular analysis. If the laboratory is not certified, and/or nonstandard methods are used, detailed quality assurance/quality control procedures must be submitted with the proposal.

Budget and Time Schedule

The maximum AwwaRF funding expected for this project is $100K. A minimum 25 percent of the total project cost must be contributed by the contractor. Therefore, the total project cost is $133K ($100K in AwwaRF funds and $33K in contractor contribution). This contribution can either be direct funding or in-kind matching of such items as personal costs, analytical and support services, facilities, consulting services, etc. The submitting organization may elect to contribute more than 25 percent to the project but AwwarF's maximum contribution remains fixed at $100K.

The project period should be realistic, anticipate possible starting delays, and provide ample time for writing, formatting, and revising of the final report and for PAC review of project results. Progress reports will generally be required on a periodic basis (every four months). The final report must be completed in accordance with the AwwaRF's *Format-Style Guide for Preparing Research Report*, unless other arrangements have been mutually agreed upon. The final report should include a separate chapter on recommendations to utilities. Independent of this contract, AwwaRF will fund the PAC as described above.

Research Project 3: Establishing the Full Costs of Injuries and Illnesses and Evaluating the Cost/Benefit of Prevention Programs

Objective

The objective of this research will be to identify the full cost of injury/illness (including cost associated with lost work time, overtime, medical costs, lost productivity, etc.) and to identify cost/benefit of developing and maintaining various types of prevention programs. This information can be used by water utility managers to make more informed decisions regarding the development of more effective safety programs.

Background

Injury and illness rates for water utilities are higher than for other utilities (i.e., power and natural gas utilities). The full economic costs associated with injuries/illnesses/fatalities in the water industry are significant but are often not accurately tracked and not very well understood in many organizations. As a result, it is sometimes difficult for safety or operations staff to justify improved safety programs.

The complete costs of injuries and illnesses goes well beyond costs tracked in workers compensation programs. Additional costs may be attributed to lost productivity, overtime for covering absent employees, medical expenses not attributed to workers compensation claims, regulatory fines, etc. Understanding the full cost of injury/illness and the cost/benefits of the various types of prevention programs will enable water systems to:

- ➢ Better justify their prevention program costs through full accounting of the benefits.
- ➢ Develop more cost-effective prevention programs
- ➢ Improve the effectiveness of intervention and rehabilitation in workers compensation cases
- ➢ Allocate their costs and prevention expenditures more appropriately between operating groups.
- ➢ Avoid running into financial surprises when serious accidents do occur.

The AWWA Health and Safety Committee has been charged with the responsibility of taking a lead role in assembling and disseminating timely state-of-the-art information to increase the health and safety of water utility workers and affected public. On its own, however, the committee cannot collect, analyze, and report accurate data on full cost of injury/illness; and evaluate the cost/benefits of prevention programs to assist water utilities in increasing the effectiveness of health and safety programs and reducing the number of injuries to their workers.

Research Approach

The research project will consist of:

- Identifying every type of costs associated with the different types of injuries and illnesses,
- Identifying the various approaches to safety management and loss prevention used by water utilities,
- Collecting data from the water utilities concerning injury/illness cases and all of the associated costs,
- Collecting data from water utilities concerning the costs of safety and loss control programs,
- Analyzing this data to estimate the full costs of injuries and illnesses and the best practices for tracking and allocating these costs,
- Evaluating costs and benefits of various types of safety management and prevention programs used by water utiilties, and
- Identifying the most cost-effective and appropriate types of prevention programs which could be implemented by water utilities to reduce injuries and illnesses to their employees.

Evaluating the cost and benefits of developing and implementing effective prevention programs will be a secondary goal of this project. The benefits of the type of prevention program to be implemented will be carefully weighed against its cost to determine type of program most appropriate for different types of water facilities.

A final project report will include:

- An explanation of the different types of costs associated with different types of injuries/illnesses,
- A summary and evaluation of data collected on various costs associated with injuries/illnesses in the water industry,
- An explanation of the different types of prevention programs,
- An evaluation of the cost/benefit of developing and implementing appropriate prevention programs,
- Providing examples of exemplary practices and the results achieved by their application, and
- Recommendations for minimizing injuries/illnesses and their associated costs.

Project Advisory Committee

Project Advisory Committees (PACs) are organized by the AWWA Research Foundation (AwwaRF) for each funded project to provide guidance, review all reports and significant materials, and generally monitor project performance in behalf of AwwaRF and the water utility industry.

Quality Assurance

Each proposal must include a description of the procedures that will be used to ensure the quality of the data for the project. If the project involves laboratory analyses, this description should indicate whether the laboratory performing the analyses is accredited or state certified for the particular analysis. If the laboratory is not certified, and/or nonstandard methods are used, detailed quality assurance/quality control procedures must be submitted with the proposal.

Budget and Time Schedule

The maximum AwwaRF funding expected for this project is $75K. A minimum 25 percent of the total project cost must be contributed by the contractor. Therefore, the total project cost is $100K ($75K in AwwaRF funds and $25K in contractor contribution). This contribution can either be direct funding or in-kind matching of such items as personal costs, analytical and support services, facilities, consulting services, etc. The submitting organization may elect to contribute more than 25 percent to the project but AwwarF's maximum contribution remains fixed at $75K.

The project period should be realistic, anticipate possible starting delays, and provide ample time for writing, formatting, and revising of the final report and for PAC review of project results. Progress reports will generally be required on a periodic basis (every four months). The final report must be completed in accordance with the AwwaRF's *Format-Style Guide for Preparing Research Report*, unless other arrangements have been mutually agreed upon. The final report should include a separate chapter on recommendations to utilities. Independent of this contract, AwwaRF will fund the PAC as described above.

Research Project 4:
National Water Occupational Health & Safety Database: Feasibility and Pilot of Benchmarking Data Consortium

Objective

The objective of this research will be to determine the feasibility of a national occupational health and safety database on injuries and illnesses in the water industry and to develop a scope for a future pilot study to test the functionality and usability of such a unified data structure.

Background

Injury and illness rates for water utilities are higher than for other utilities (i.e., power and natural gas utilities) based upon data compiled by the Bureau of Labor Statistics (BLS) for privately operated systems. OSHA evaluations have suggested that, in general, injury and illness rates for government workers not covered under Federal OSHA reporting requirements are higher than for private industry. OSHA and BLS agree that there are currently inadequate data collection mechanisms to fully assess the performance of publicly operated water systems because they are not subject to Federal OSHA regulations, including reporting requirements. As a result, it is difficult for water utilities to benchmark themselves against the performance of similar operations or different utility/industry sectors to determine whether they require improved safety programs.

It is important for water systems to benchmark their performance against others in their industry and other industry sectors. Benchmarking enables water systems to determine whether their performance warrants improvement. Well designed benchmarking tools can allow them to go much further and identify critical areas experiencing higher incidence rates that need to be focused on for incidence reduction. However, the water industry does not have a standard format for collecting and reporting incidence/injury data.

There is a critical need for uniform reporting by all water utilities. There would be even greater benefits for a unified data structure that encompasses all the current data reporting requirements (e.g. regulatory, insurance, workers' compensation, etc.) into one and provides a more detailed breakdown of performance within an organization (e.g., provides for breakdown between departments and workgroups and for recording types and root causes of incidents). Such a system would allow water systems to not only generate required reports but also conduct detailed analysis of incidents within the organization and compare their safety performance with others.

The AWWA Health and Safety Committee has been charged with the responsibility of taking a lead role in assembling and disseminating timely state-of-the-art information to increase the health and safety of water utility workers and affected public. On its own, however, the committee cannot undertake a project of this magnitude and complexity. Some of the major steps to be taken by this program include: collecting data from the water industry, developing data structure encompassing existing databases, conducting a pilot study, and making the data available the data online in secure manner.

Research Approach

This research will be broken up into two phases. The first phase, to be conducted in this project, will determine the feasibility and interest in creating a national incident/injury/illness database and develop a unified data structure for the database. This research project should include key stakeholders (e.g. water industry, insurance companies, workman compensations, researchers, and federal and private research organizations) whose input will provide valuable insight in developing this database. The second phase will then be to conduct a pilot study to test the functionality, and usability of this data structure.

The Feasibility Study phase will need to:

- Determine the full range of injury/illness data that must be collected to meet government, insurance and workers compensation program requirements.
- Determine the data collected in exemplary incident reporting and analysis systems in the water industry and other related industries to allow thorough analysis of areas requiring improvement.
- Identify relevant organizational characteristics (e.g., private/public, water/combined utilities, regulatory environment, size, unionization, etc.) that allows a more thorough benchmarking of performance.
- Develop a unified data structure that encompasses both input/output of required information and information that supports robust analysis, benchmarking and corrective action.
- Develop methods for ensuring the reliability and validity of data.
- Identify the interest in using such a database as the primary system for external reporting, internal performance analysis and benchmarking.
- Determine cost of developing and maintaining the database.
- Identify marketability of the database and evaluate options for recovering the cost of implementation (including membership fees).
- Identify the feasibility of making this database internet-based (for greater availability to water systems).
- Evaluate integrating this database with other existing databases such as Water Stats, and Qualserve

A final project report will include:

- A description of the methods and results of the feasibility study.
- A unified data structure that encompasses both input/output of required information and information that supports robust analysis, benchmarking and corrective action
- Analysis of implementation options, costs, and marketability of the unified database.
- Recommendations on whether to proceed to the pilot study phase and a scope and budget for the pilot study.

Project Advisory Committee

Project Advisory Committees (PACs) are organized by the AWWA Research Foundation (AwwaRF) for each funded project to provide guidance, review all reports and significant materials, and generally monitor project performance in behalf of AwwaRF and the water utility industry.

Quality Assurance

Each proposal must include a description of the procedures that will be used to ensure the quality of the data for the project. If the project involves laboratory analyses, this description should indicate whether the laboratory performing the analyses is accredited or state certified for the particular analysis. If the laboratory is not certified, and/or nonstandard methods are used, detailed quality assurance/quality control procedures must be submitted with the proposal.

Budget and Time Schedule

The maximum AwwaRF funding expected for this project is $150K. A minimum 25 percent of the total project cost must be contributed by the contractor. Therefore, the total project cost is $200K ($150K in AwwaRF funds and $50K in contractor contribution). This contribution can either be direct funding or in-kind matching of such items as personal costs, analytical and support services, facilities, consulting services, etc. The submitting organization may elect to contribute more than 25 percent to the project but AwwarF's maximum contribution remains fixed at $150K. {Note: a preliminary estimate of the subsequent pilot study phase is $250K to $300K in AwwaRF funding.}

The project period should be realistic, anticipate possible starting delays, and provide ample time for writing, formatting, and revising of the final report and for PAC review of project results. Progress reports will generally be required on a periodic basis (every four months). The final report must be completed in accordance with the AwwaRF's *Format-Style Guide for Preparing Research Report*, unless other arrangements have been mutually agreed upon. The final report should include a separate chapter on recommendations to utilities. Independent of this contract, AwwaRF will fund the PAC as described above.

Research Project 5: Evaluation of the Effectiveness of Alternate Enforcement Methods and Approaches

Objective

The objective of this research will be to identify and evaluate the effectiveness of alternate enforcement methods in improving water utility safety performance.

Background

Injury and illness rates for water utilities are higher than for other utilities (i.e., power and natural gas utilities). There are many reasons why water systems should endeavor to create a more safe and healthful workplace, ranging from concern about their employees to controlling costs to preventing enforcement action by OSHA or state occupational health and safety agencies.

Many water systems adopt programs and procedures and maintain a strong safety culture that go well beyond that required by law; however, all too often regulatory enforcement is the only factor driving health and safety programs. In fact, some water industry safety representatives report that they must sometimes resort to asking enforcement representatives to "write them a ticket" to get problems addressed.

Traditional enforcement is a weak tool in the water industry for the following reasons:

- OSHA and many state enforcement agencies have a limited enforcement budget, which limits enforcement activities and therefore makes enforcement a weak tool for performance improvement.
- Government operated systems are not subject to Federal OSHA regulations and only about half of the states have equivalent state requirements.
- Water systems are not widely perceived as a "high risk" industry and do not receive much enforcement attention.

Some organizations, including water systems, take advantage of limited enforcement by adopting a philosophy of waiting to be cited. For instance, water systems might weigh the cost of developing an effective training program against the cost of being fined by OSHA and conclude that it is more cost efficient to risk the fine instead. Many water systems in states without OSHA-type rules have voluntary programs that do not provide comprehensive prevention measures comparable to OSHA-type programs. Such an approach is often taken at the risk of endangering the health and safety of employees.

Alternative enforcement methods and approaches to encourage voluntary action have been developed by some states and by U.S. OSHA over the last few years. Some examples include New Jersey's empowering State Troopers to issue citations over trenching violations and OSHA's Voluntary Protection Program (VPP) that encourages safety excellence. It would be beneficial to establish what alternative programs exist and to evaluate how effective they are in making industry in general, and specifically water systems, adopt practices to promote the safety and health of their workers.

The AWWA Health and Safety Committee has been charged with the responsibility of taking a lead role in assembling and disseminating timely state-of-the-art information to increase the health and safety of water utility workers and affected public. On its own, however, the committee cannot collect, analyze, and report accurate information on the effectiveness of alternative enforcement methods to assist water utilities in increasing the effectiveness of health and safety programs and reducing the number of injuries to their workers.

Research Approach

The research will identify the different enforcement approaches and measures of the rigor of enforcement (e.g., the amount of resources applied by OSHA and various state occupational safety and health agencies) and evaluate whether there is any correlation between rigorous enforcement and safety performance. It will also identify alternate enforcement methods/approaches by conducting literature searches, and obtaining input from federal and state occupational health and safety and workers compensation agencies, and water systems throughout the country.

A final project summary will include:

- A review of current conventional enforcement methods,
- An evaluation of the relationship between rigorous enforcement and safety performance,
- Identification of areas where current methods fail short
- A review of alternate enforcement programs describing who is the enforcement agency (state police, fire department etc.) and how the program works.
- An evaluation of the effectiveness of alternate enforcement methods
- Recommendations on improve current enforcement methods and on which alternative enforcement approaches should be considered for wider use.

Project Advisory Committee

Project Advisory Committees (PACs) are organized by the AWWA Research Foundation (AwwaRF) for each funded project to provide guidance, review all reports and significant materials, and generally monitor project performance in behalf of AwwaRF and the water utility industry.

Quality Assurance

Each proposal must include a description of the procedures that will be used to ensure the quality of the data for the project. If the project involves laboratory analyses, this description should indicate whether the laboratory performing the analyses is accredited or state certified for the particular analysis. If the laboratory is not certified, and/or nonstandard methods are used, detailed quality assurance/quality control procedures must be submitted with the proposal.

Budget and Time Schedule

The maximum AwwaRF funding expected for this project is $75K. A minimum 25 percent of the total project cost must be contributed by the contractor. Therefore, the total project cost is $100K ($75K in AwwaRF funds and $25K in contractor contribution). This contribution can either be direct funding or in-kind matching of such items as personal costs, analytical and support services, facilities, consulting services, etc. The submitting organization may elect to contribute more than 25 percent to the project but AwwarF's maximum contribution remains fixed at $75K.

The project period should be realistic, anticipate possible starting delays, and provide ample time for writing, formatting, and revising of the final report and for PAC review of project results. Progress reports will generally be required on a periodic basis (every four months). The final report must be completed in accordance with the AwwaRF's *Format-Style Guide for Preparing Research Report*, unless other arrangements have been mutually agreed upon. The final report should include a separate chapter on recommendations to utilities. Independent of this contract, AwwaRF will fund the PAC as described above.

Research Project 6:
Effective Infrastructure Protection Techniques for Protection of Water System Employees and the Public

Objective

The objective of this research will be to identify and evaluate effective infrastructure protection techniques in preventing injury to water system employees and the public.

Background

Of the seven critical infrastructure sectors identified by the U.S. government, water supply is the only one that involves direct consumption in the human body. As such, it is not only important but often there is an emotional aspect to the consumer's perceptions of their water supply.

Both the U.S. government and the water industry recognized the need to enhance current protections of our nation's infrastructure long before the events of September 11th, 2001. Since then, many water utilities have taken extensive measures for upgrading security. In addition, some water utilities have specifically developed techniques and fabricated equipment for water industry applications (e.g., special reservoir vent hoods). The long-term effectiveness of these measures in reducing risk has not been fully evaluated.

The vast majority of the security measures in the water industry have been taken at fixed facilities, but significant vulnerabilities exist elsewhere in the system. While water systems have raised legitimate questions about the efficacy and costs of measures to harden the distribution system, there is not a wide awareness of the inherently safer design and equipment options that are currently available.

The concept of inherently safer system design is founded on the understanding that the best opportunity for integrating inherent protections into water treatment plants, conveyance systems, storage and distribution systems is at the design stage. The earlier in the design, the greater the improvements that can be made and the lower the cost. In addition to currently available options, there may be potential new inherently safer design approaches and equipment that warrant development and demonstration.

Accordingly, there is a critical need for:

- Reliable information resources for water utility managers regarding the availability, cost, and use in the industry of protection and prevention techniques and devices.
- Identifying and communicating exemplary practices to improve infrastructure protection during the design and operation of water systems. A good understanding of exemplary practices will enable designers and owners of other water systems to benchmark and improve their own approaches to designing and operating their systems.
- Identifying potential new inherently safer methods that warrant further development and evaluation.

Research Approach

The research will identify and evaluate existing design and operations practices and equipment for water system infrastructure protection (e.g., physical facility intrusion detection and response, water contaminant detection, distribution system hardening, etc.), including novel techniques and exemplary practices. Methods may include conducting a survey to collect further information on successful approaches, and conducting a workshop for identifying best practices to gain input from experts of the various stakeholders for security in the water industry.

The research will also identify critical system design, operations and equipment needs that are not being met by currently available techniques and equipment.

The final project report will include a description of currently available resources and exemplary practices in the water industry for the design and operation of water systems, an evaluation of their effectiveness and advice on how to develop and implement these practices at water utilities. The report may also include examples of lessons learned in these organizations through past experiences that led to the development of the exemplary practices and recommendations on design and operation approaches to enhance inherent system security.

A final project summary will include:

- Effective infrastructure protection techniques and an evaluation of their effectiveness,
- Examples of exemplary infrastructure protection practices in the water industry,
- Infrastructure protection techniques that do not work (lessons learned through past mistakes),
- An accounting of currently unmet needs for infrastructure protection, and
- Recommendations for development and testing of new equipment or design approaches that can meet these unmet needs.

Project Advisory Committee

Project Advisory Committees (PACs) are organized by the AWWA Research Foundation (AwwaRF) for each funded project to provide guidance, review all reports and significant materials, and generally monitor project performance in behalf of AwwaRF and the water utility industry.

Quality Assurance

Each proposal must include a description of the procedures that will be used to ensure the quality of the data for the project. If the project involves laboratory analyses, this description should indicate whether the laboratory performing the analyses is accredited or state certified for the particular analysis. If the laboratory is not certified, and/or nonstandard methods are used, detailed quality assurance/quality control procedures must be submitted with the proposal.

Budget and Time Schedule

The maximum AwwaRF funding expected for this project is $150K. A minimum 25 percent of the total project cost must be contributed by the contractor. Therefore, the total project cost is

$200K ($150K in AwwaRF funds and $50K in contractor contribution). This contribution can either be direct funding or in-kind matching of such items as personal costs, analytical and support services, facilities, consulting services, etc. The submitting organization may elect to contribute more than 25 percent to the project but AwwarF's maximum contribution remains fixed at $150K.

The project period should be realistic, anticipate possible starting delays, and provide ample time for writing, formatting, and revising of the final report and for PAC review of project results. Progress reports will generally be required on a periodic basis (every four months). The final report must be completed in accordance with the AwwaRF's *Format-Style Guide for Preparing Research Report*, unless other arrangements have been mutually agreed upon. The final report should include a separate chapter on recommendations to utilities. Independent of this contract, AwwaRF will fund the PAC as described above.

Research Project 7: The Role of Human Factors in Injuries and Improving Safety Performance in the Water Industry

Objective

The objective of this research will be to better understand the effects of human factors on safety performance and to provide guidance to improve the health and safety of water utility workers and prevent potential effects to the public by considering human factors in system design and operations.

Background

Human factors have been extensively studied as they pertain to potential accidents in space travel, aviation, and nuclear power plant operation. Consideration of human factors is increasingly cited as a recommended practice (e.g., by the Center for Chemical Process Safety) and as a regulatory requirement (e.g., in EPA Risk Management Programs regulations, 40 CFR Part 68). As a result, human factors are increasingly being considered in hazard analysis, post-incident investigation, in design of new facilities and in developing standard operating and maintenance procedures.

Injury and illness rates for water utilities are higher than for other utilities (i.e., power and natural gas utilities). It is well-established that human factors are important root causes and contributing causes of most accidents, but human factors issues specific to the water industry have not been studied.

Human factors, as practiced in the safety management field, is a multidisciplinary effort to collect and evaluate information about human capabilities and limitations and apply that information to equipment, systems, software, facilities, procedures, jobs, environments, training, staffing, and personnel management to produce safe, comfortable, and effective human performance.

When human factors is applied early in the process of designing a plant and developing operating procedures, it:

- Improves performance, safety, and productivity;
- Decreases lifecycle staffing and training costs; and
- Becomes integrated into business strategy and planning.

Changes in operational, maintenance or design concepts during the later phases of a project are expensive and entail high-risk program adjustments. Identifying lifecycle costs and human performance components of system operation and maintenance during project definition decreases risks and long-term operations costs.

Water utility safety managers have noted that they believe human factors such as employee experience, employee education, age of employee, etc. can affect safety performance.

Understanding these and other human factors is essential in improving safety performance. Some of issues that need to be better characterized are:

- Safety Performance vs. Work Environment Factors: Are employees more likely to get injured working alone? How important is language/communication between employees?
- Safety Performance vs. Design of Systems, Job Tools and Methods: Do active hazard analysis and developing systems, procedures and tools to safely perform the job task reduce injuries (e.g., does ergonomic plant design or field tools such as a hoist to remove piping/valves from an excavation reduce injuries)?
- Safety Performance vs. Stress: Are high stress jobs more likely to result in more accidents (e.g. overtime, cross-functional responsibilities, multiple facility responsibilities, working alone)?
- Safety Performance vs. Safety Culture: How are employees affected by their safety culture?
- Safety Performance vs. Age of Employee: Are there some demanding jobs that are more likely to cause injury to all populations because the methods used are near the limits of human capabilities? Are older employees more or less likely to get these types of injuries? Do younger employees recover from their injuries more quickly?
- Safety Performance vs. Experience/Length of Service: Some water industry data suggest that new employees have fewer injuries than those with 5 or more years of experience. What is the relationship of incidence rates to experience and what are the causal factors for these differences (e.g., do employees tend to adhere more strictly to safety guidelines the longer they are in the business, or do they get more comfortable with their work and begin cut corners or be more likely to make errors).
- Safety Performance vs. History of Injury/Illness: Are employees with a history of injury or health problems more or less likely to get seriously injured?

The AWWA Health and Safety Committee has been charged with the responsibility of taking a lead role in assembling and disseminating timely state-of-the-art information to increase the health and safety of water utility workers and affected public. An annual Health and Safety survey collects some data but does not include collecting, analyzing, and reporting statistical data on human factors affecting safety performance to assist water utilities in increasing the health and safety of their workers.

Research Approach

The research will include collecting and evaluating data on the effect of human factors on injuries and illnesses in the water industry. The project will focus on comparing the frequency and severity of injuries of employees based on such factors as:

- Type of work
- Years of work experience
- Age group
- Education level
- Time of day (shift)
- Working alone

- Language spoken
- Length of workday

The study will also identify and evaluate exemplary practices of applying human factors to the design of plants and to the procedures and tools used in field construction/operations/maintenance job tasks (especially those activities that have higher injury rates). A final project summary will include:

- Information on various human factors and how they effect safety performance.
- Description of best practices in applying human factors in plant design and in performing construction, operation and maintenance job tasks.
- Guidelines for how to utilize this understanding of human factors to improve safety performance.

Project Advisory Committee

Project Advisory Committees (PACs) are organized by the AWWA Research Foundation (AwwaRF) for each funded project to provide guidance, review all reports and significant materials, and generally monitor project performance in behalf of AwwaRF and the water utility industry.

Quality Assurance

Each proposal must include a description of the procedures that will be used to ensure the quality of the data for the project. If the project involves laboratory analyses, this description should indicate whether the laboratory performing the analyses is accredited or state certified for the particular analysis. If the laboratory is not certified, and/or nonstandard methods are used, detailed quality assurance/quality control procedures must be submitted with the proposal.

Budget and Time Schedule

The maximum AwwaRF funding expected for this project is $200K. A minimum 25 percent of the total project cost must be contributed by the contractor. Therefore, the total project cost is $250K ($200K in AwwaRF funds and $50K in contractor contribution). This contribution can either be direct funding or in-kind matching of such items as personal costs, analytical and support services, facilities, consulting services, etc. The submitting organization may elect to contribute more than 25 percent to the project but AwwarF's maximum contribution remains fixed at $200K.

The project period should be realistic, anticipate possible starting delays, and provide ample time for writing, formatting, and revising of the final report and for PAC review of project results. Progress reports will generally be required on a periodic basis (every four months). The final report must be completed in accordance with the AwwaRF's *Format-Style Guide for Preparing Research Report*, unless other arrangements have been mutually agreed upon. The final report should include a separate chapter on recommendations to utilities. Independent of this contract, AwwaRF will fund the PAC as described above.

Research Project 8: The Effects of Chronic Low Level Exposures to Airborne Chemicals at Water Utilities

Objective

The objective of this research will be to identify and understand the extent and effects of chronic low level exposure to chlorine, ammonia and/or chloramine vapors by employees in the water industry and to provide guidance on how to minimize/prevent employee exposures to these vapors.

Background

Workers in the water industry are sometimes exposed to airborne chemicals, most often chlorine, ammonia and chloramine vapors from disinfection (i.e., from the use of chlorine, ammonia and sodium hypochlorite). While effects of acute exposure to these chemicals are easily identifiable, the extent of exposure of water treatment plant employees and potential health effects due to chronic low-level exposures are not very well understood. Health effects become even more difficult to identify and treat when synergistic effects of exposure to chronic low-level chlorine and exposure to other chemical and/or physical agents are considered. New approaches are needed to identify and understand chronic low-level exposures and their effects. Depending upon the results, it may be necessary to provide guidance on improved hazard controls to minimize/prevent future employee exposure and/or to better assess and treat affected employees.

The AWWA Health and Safety Committee has been charged with the responsibility of taking a lead role in assembling and disseminating timely state-of-the-art information to increase the health and safety of water utility workers and affected public. On its own, however, the committee cannot collect, analyze, and report statistically significant data on airborne chemical exposures and related injury/illness; evaluate the health effects; identify techniques for preventing potential problems; and provide guidance on effective prevention techniques.

Research Approach

Research on low-level chronic chlorine/ammonia/chloramine vapors exposure will include collecting chronic exposure data, assessing the health effects based on current scientific and medical information and studies, and making appropriate recommendations for exposure assessment and control in the design, operations and maintenance of water treatment plants.

The exposure assessment will consist of collecting and analyzing information gathered from the water industry. It may include searching for existing exposure assessment data from water utilities and published studies, but it will also include identifying a representative bounding set of water treatment plants and conducting monitoring for exposure assessment of employees working with chlorine/hypochlorite/ammonia. The assessment will also identify the characteristics of the facilities and operation and maintenance practices that might affect the magnitude of these low-level exposures. Water utilities will be surveyed to determine whether any maintain records that could be used to retrospectively assess health effects of treatment plant workers vs. individuals who do not work with these chemicals.

The health effects assessment will consist of gathering scientific and medical information to understand the known health effects of chronic low-level chlorine/ammonia/chloramine exposure. This may include literature review, collaborating with researchers in the field or conducting a workshop. Available information from sources such as molecular biology and epidemiology research will be used to assess the potential effects of low level chronic worker exposure.

Based upon the relationship of the facility/operations characteristics to the exposure assessment, as well as basic principles of industrial ventilation, conclusions should be developed about appropriate design and operations practices that can be put into the form of recommendations for exposure assessment and control in the design, operations and maintenance of water treatment plants.

A final project summary will include:

- Results of the chronic exposure assessment,
- Results of the health effects assessment based on current scientific and medical information and studies,
- Assessment of the design and operations factors that may be affecting concentrations in the work place, and
- Recommendations for conducting workplace exposure assessment and for applying good control practices in the design, operations and maintenance of water treatment plants to minimize/prevent employee exposure.

Project Advisory Committee

Project Advisory Committees (PACs) are organized by the AWWA Research Foundation (AwwaRF) for each funded project to provide guidance, review all reports and significant materials, and generally monitor project performance in behalf of AwwaRF and the water utility industry.

Quality Assurance

Each proposal must include a description of the procedures that will be used to ensure the quality of the data for the project. If the project involves laboratory analyses, this description should indicate whether the laboratory performing the analyses is accredited or state certified for the particular analysis. If the laboratory is not certified, and/or nonstandard methods are used, detailed quality assurance/quality control procedures must be submitted with the proposal.

Budget and Time Schedule

The maximum AwwaRF funding expected for this project is \$225K. A minimum 25 percent of the total project cost must be contributed by the contractor. Therefore, the total project cost is \$300K (\$225K in AwwaRF funds and \$75K in contractor contribution). This contribution can either be direct funding or in-kind matching of such items as personal costs, analytical and support services, facilities, consulting services, etc. The submitting organization may elect to contribute more than 25 percent to the project but AwwarF's maximum contribution remains fixed at $225K.

The project period should be realistic, anticipate possible starting delays, and provide ample time for writing, formatting, and revising of the final report and for PAC review of project results. Progress reports will generally be required on a periodic basis (every four months). The final report must be completed in accordance with the AwwaRF's *Format-Style Guide for Preparing Research Report*, unless other arrangements have been mutually agreed upon. The final report should include a separate chapter on recommendations to utilities. Independent of this contract, AwwaRF will fund the PAC as described above.

REFERENCES

Almond. Jeff. 2001. Best Practices for Power Industry Data Collection/Analysis. Presented at AwwaRF Workshop for Identifying and Prioritizing Emerging Safety Issues in the Water Industry, Ft. Lauderdale, Florida, October 28-30, 2001.

AWWA (American Water Works Association). 1999. AWWA Safety Survey for Water Utilities.

Brown, Randy. 2001. Key Elements of an Effective Safety Program. Presented at AwwaRF Workshop for Identifying and Prioritizing Emerging Safety Issues in the Water Industry, Ft. Lauderdale, Florida, October 28-30, 2001.

BLS (Bureau of Labor Statistics). 2000. Occupational Injuries and Illnesses in the United States, Profile Data 1992-2000.

Hanley, Johnathan. 2001. Water Supply - Safety and Health: A Regulatory Perspective. Presented at AwwaRF Workshop for Identifying and Prioritizing Emerging Safety Issues in the Water Industry, Ft. Lauderdale, Florida, October 28-30, 2001.

Hoskin, Alan F. 2001. Traumatic Occupational Injury Research Needs and Priorities. Presented at AwwaRF Workshop for Identifying and Prioritizing Emerging Safety Issues in the Water Industry, Ft. Lauderdale, Florida, October 28-30, 2001.

Markiewicz, Bobbi, PhD. 2001. System Outcome Measures for Workplace Disability Prevention. Presented at AwwaRF Workshop for Identifying and Prioritizing Emerging Safety Issues in the Water Industry, Ft. Lauderdale, Florida, October 28-30, 2001.

NSC (National Safety Council). 2000. *National Safety Council Injury Facts.* Itasca, IL: Author.

Puglionesi, Peter. 2001. Evaluation of Current Water Industry Injury/Illness Data. Presented at AwwaRF Workshop for Identifying and Prioritizing Emerging Safety Issues in the Water Industry, Ft. Lauderdale, Florida, October 28-30, 2001.

Skidmore, Ben. 2001. Behavior-Based Employee-Driven Safety Program. Presented at AwwaRF Workshop for Identifying and Prioritizing Emerging Safety Issues in the Water Industry, Ft. Lauderdale, Florida, October 28-30, 2001.

U.S. Department of Labor. 2000. *Evaluating the Status of Occupational Safety and Health Coverage of State and Local Government Workers in Federal OSHA States.* (Report Number 05-00-0001-10-001).

ABBREVIATIONS

ANSI	American National Standards Institute
AWWA	American Water Works Association
AwwaRF	Awwa Research Foundation
BLS	Bureau of Labor Statistics
EEI	Edison Electric Institute
EH&S	Environmental Health and Safety
EPA	Environmental Protection Agency
EPRI	Electric Power Research Institute
F-DE&S	Framatome ANP Duke Engineering and Services
H&S	Health and Safety
IAIABC	International Association of Industrial Accident Boards and Commissions
ISO	International Organization for Standardization
MSD	Musculo Skeletal Disorder
NIOSH	National Institute for Occupational Safety and Health
NSC	National Safety Council
NY	New York
OSHA	Occupational Safety and Health Administration
PAC	Project Advisory Committee
PESH	Public Employees Safety and Health
R&D	Research and Development
SAWS	San Antonio Water System
U.S.	United States
VPP	Voluntary Protection Program
WEF	Water Environment Federation
WERF	Water Environment Research Federation
WTP	Water Treatment Plant

IWA Publishing's authorised EU representative for General Product Safety Regulations is Diane D'Arras, 15 rue Duret, 75116 Paris, France, e-mail: safety@iwap.co.uk.

Printed and bound by CPI Group (UK) Ltd, Croydon, CR0 4YY
17/09/2026
02218813-0012